Guide to Complete Streets Campaigns

Guide to Complete Streets Campaigns

◆

**For Thunderhead Alliance
Member Organizations
March 2006**

As part of Thunderhead's 50 States/50 Cities Project

*Produced by Thunderhead Alliance
Sue Knaup, Executive Director
Gayle Stallings
David Crites
Dave Snyder
and
Barbara McCann, McCann Consulting*

iUniverse, Inc.
New York Lincoln Shanghai

Thunderhead Alliance Guide to Complete Streets Campaigns
For Thunderhead Alliance Member Organizations March 2006

iUniverse books may be ordered through booksellers or by contacting:

iUniverse
2021 Pine Lake Road, Suite 100
Lincoln, NE 68512
www.iuniverse.com
1-800-Authors (1-800-288-4677)

First Edition: March 2005

ISBN-13: 978-0-595-39318-3 (pbk)
ISBN-13: 978-0-595-83714-4 (ebk)
ISBN-10: 0-595-39318-7 (pbk)
ISBN-10: 0-595-83714-X (ebk)

Printed in the United States of America

With funding assistance from Planet Bike
and the National Bicycle Dealers Association

Contents

ACKNOWLEDGEMENTS

The Thunderhead Alliance's <u>Guide To Complete Streets Campaigns</u> was made possible with the responses from bicycle and pedestrian advocacy organization leaders, bicycle/pedestrian coordinators, and other public service staff from across the country. They took the time to fill out our complete streets survey and/or responded to telephone calls about their experiences with complete streets type policies and campaigns. Thanks to these entities:

- Arizona Department of Transportation
- Bicycling for Louisville
- BicycleAccess-PA
- BikeWalk Virginia
- Bicycle Transportation Alliance
- California Department of Transportation
- Central Ohio Bicycle Advocacy Coalition
- Chris Morfas, Odyssey
- City of Boulder, Colorado
- City of DuPage, Illinois
- City of Fort Collins, Colorado
- City of Portland, Oregon
- City of Sacramento, California
- City of Santa Barbara, California
- City of St. Joseph, Missouri
- City of St. Petersburg, Florida
- City of West Palm Beach, Florida
- Florida Department of Transportation

- Knoxville Regional Transportation Planning Organization
- League of Illinois Bicyclists
- Marin County Bicycle Coalition
- North Carolina State University Institute for Transportation Research and Education
- Northeast Ohio Areawide Coordinating Agency
- Oregon Department of Transportation
- Pednet
- Rhode Island Department of Transportation
- Sacramento Area Bicycle Advocates
- SANDAG
- San Diego County Bicycle Coalition
- South Carolina Department of Transportation
- Tennessee Department of Transportation
- Virginia Department of Transportation

Sincere thanks go out to the executive directors and staff at those organizations that we singled out for campaign examples. They patiently took our questions and phone calls and even proofed to make sure that we got it right. These are:

- Ed Barsotti, League of Illinois Bicyclists
- John Gideon, Central Ohio Bicycle Advocacy Coalition
- Deb Hubsmith, Marin County Bicycle Coalition
- Robin Stallings, Texas Bicycle Coalition

Essential guidance and input came from the Thunderhead Complete Streets Committee comprised of Randy Neufeld, Barb Culp, Dan Grunig, Noah Budnick, Eric Gilliland and Sue Knaup.

Thanks also go out to Nancy Weaver, who assisted with some of the campaign research, writing, and documentation, and the Sierra Club, who have wonderful publications related to environmental issues and grassroots campaigns that were an inspiration to us. One such publication, How to Guide: Grassroots Organizing on Texas Water Issues, produced by the Lone Star Chapter of the Sierra Club, provided some of the framework in creating Thunderhead's Campaign Planning Blueprint.

Of course, this Guide to Complete Streets Campaigns would not have been possible without the generous financial assistance from Planet Bike, the National Bicycle Dealers Association and the

New-Land Foundation who believe in the power of effective change through the Thunderhead Alliance's 50 States/50 Cities Project.

If we forgot anyone, please forgive us and let us know. Then watch for your name in our next update.

Sue Knaup   Gayle Stallings   David Crites   Dave Snyder   Barbara McCann

1

Introduction

Thunderhead Alliance

The Thunderhead Alliance is the national coalition of state and local bicycle and pedestrian advocacy organizations. 119 strong in 48 states and one Canadian province, Thunderhead's member organizations employ more than 187 full-time staff and reach a combined dues-paying membership of more than 94,000 people. Thunderhead's mission is to unite these organizations, help strengthen them, and create new ones where they are most needed.

Complete streets policies that require safe accommodation of all users of a street can eliminate most of this nation's barriers to bicycling and walking. Thunderhead's National Complete the Streets Campaign has a goal of helping our organizations win at least one complete streets policy, local or state-level, in all 50 states by 2008 in order to influence a federal-level complete streets policy through the reauthorization of SAFETEA-LU, the U.S. federal transportation law. This tapestry of local, state and federal policies will ensure that no transportation project can move forward without being complete!

If you are a leader or potential leader of a Thunderhead organization, this <u>Guide to Complete Streets Campaigns</u> is written for you. If you are not a leader of such an organization, this Guide will be your window into the world of bringing positive change to communities through professional bicycle and pedestrian advocacy. Read as if you are a leader of a Thunderhead organization and bring these elements of this powerful transportation reform campaign to your own officials. Sometimes all it takes is one determined, professional voice. And make sure to connect with your Thunderhead organization on our Links page at: www.thunderheadalliance.org/links.htm .

Thunderhead's National Complete the Streets Campaign is at the heart of Thunderhead's 50 States/50 Cities Project, which has a goal of effective and sustainable bicycle and pedestrian advocacy organizations in all 50 states and at least the 50 top population U.S. cities. The 50 State/50 Cities Project thrives due in large part to the significant, farsighted support of Planet Bike and the National Bicycle Dealers Association.

This <u>Guide to Complete Streets Campaigns</u> is a roadmap to winning a complete streets policy in your jurisdiction. It is also a guide to effective community organizing, as it is our hope that in winning a complete streets policy our Thunderhead member organizations will also gain strength, increase partnerships, and in many ways make their communities better with improved conditions for bicycling and walking.

Please note that this document cannot substitute for Thunderhead's on-call assistance or in person campaign trainings. Thunderhead staff are available at all times to assist the leaders of our member organizations in organizational development and campaign issues. We can help you find your unique path through this document to create a campaign that is just right for your unique situation. And once you've launched your campaign, we are here to help you at all steps along the way. Contact us anytime at our Main Office at: 928-541-9841 or at our Washington, DC Office: 202-349-1479.

The Concepts of Complete Streets and Complete the Streets Campaigns

Complete streets are thoroughfares that serve all users, moving by car, truck, transit, bicycle, wheelchair, or foot. Complete streets allow all their users to travel in a safe and welcoming way. You, as a leader of a Thunderhead organization, as a champion of bicycling and walking issues, as a bicyclist and pedestrian, will acknowledge that the vast majority of the current North American transportation system is not comprised of complete streets. Many streets lack sidewalks, few accommodate bicyclists well, most encourage traffic to travel too close and fast, many don't have curb ramps at intersections or across driveways, and so on. We all know that these types of streets are less safe, less functional, and a hindrance to healthy communities and people.

The Cost Misconception: A common misconception is that complete streets cost more to build than incomplete streets. In fact, complete streets most often cost no more and many times can cost less than incomplete streets. For instance, a common street cross section that serves only cars is a four lane speedway with no shoulders, sidewalks or intersection treatments for people. Using the same right-of-way width, this design can be reshaped into two narrower through lanes, one center turn lane, and bike lanes and sidewalks on both sides. By using less width for the most expense elements, truck weight standard asphalt and subsurface, and adding less expensive sidewalks, this design, often referred to as a "road diet" when applied to existing roads, actually saves money. Not only that, this design has been proven to improve traffic flow and safety for motor vehicles by better controlling turning movements. Many other complete streets designs offer similar cost savings. You may even want to bring up the economic benefits of streets that attract visitors and offer access to more employees. Be sure to address this misconception early in your campaign so that you can focus your valuable time on instituting a policy for your communities.

Complete streets is a newly coined phrase meant to take the older concept of "routine accommodation" to the next level. Doesn't it just sound better to you? The phrase is action-oriented and flexible. For example, imagine saying to a reporter that "we are completing the streets" or saying it mantra-style at a rally: "complete the streets, complete the streets, complete the streets." The phrase also conveys an important underlying message—that streets are not complete until they are designed, built or rebuilt, and operated in a manner that provides safe and reasonable travel for all modes. A street that does not provide for such passage is, by default, incomplete. The term complete streets is already popular in the United States; using this term will make it easier for you to emphasize the need for bicycle and pedestrian safety to elected officials, agency staff and community leaders alike.

Winning a complete streets policy and getting it implemented properly will benefit you in many ways. At the top of the list is the promise that you don't have to fight for each and every street to be constructed or reconstructed appropriately and completely. Your organization can spend more time and  resources on activities other than chasing every project. Your organization will reap rewards for having done something proactive for your members and for your community. Your organization may also benefit from more partnerships due to the broad appeal of a complete streets campaign. It could gain clout too, if the organization is seen as capable of pulling off a big campaign in partnership with other organizations and decision makers. Finally and a bit selfishly, you and your loved ones will gain since you are active participants in the transportation system.

All that said, we currently have neither the definitive complete streets policy nor a simple "rubber stamp" campaign that will work in every community. Our extensive survey of complete streets type policies and outreach have so far yielded 24 model policies across the United States, but they are wide ranging in type and degree. The majority of them are only a few years old with short or no track records. We also discovered 12 "paper" policies that are not quite models but include good starting elements. See Chapter 2 for the complete list.

On the campaign side, prior to Thunderhead's launch of our National Complete the Streets Campaign in 2004, there were few campaigns that sought policies that would result in complete streets. However, there is plenty of history regarding successful advocacy campaigns. This Guide relates that history to campaigns for complete streets policies. We look forward to future updates of this Guide that will include a variety of powerful models for comprehensive complete streets policy campaigns. The first insight from successful campaigns is that your own knowledge of local and state politics, legislative processes, transportation agencies, communities, resources, and etceteras, will determine your best course of action.

Your own circumstances will best determine what kind of complete streets policy to pursue. There are at least three types: legislative requirements of transportation agencies, internal transportation agency policies, and design guidelines that require complete streets. You may find that there is already a complete streets policy in place. If so, the policy may just need some implementation assistance or refinement to make it a more effective, successful policy. If not, ask your transportation officials about the possibility of getting a policy. A sympathetic and influential few could negate the need for a difficult campaign. At least you will get valuable information from these officials that could make your complete streets campaign a row much easier to hoe.

Be wary however, of "poser" policies or policies that staff are never trained on and/or that are never implemented. Such poser policies make an agency or jurisdiction look good, but do little to improve conditions for all users of a thoroughfare or transportation network sometimes even blocking these provisions. One of the most important findings of Thunderhead member organizations through the years has been that policies can get approved, but without leadership, follow-up, and/or training, they will never hit the streets. Ensuring they hit the streets is one of the most important roles of Thunderhead organizations. Be aware that some related policies, like context sensitive design, may even be used to the detriment of bicycling and walking.

Should you end up pursuing a complete streets campaign, remember that if you do get bogged down, try not to get stuck with nothing at the end. Get what you can, pass what you can pass, and work to make it better during the implementation processes or by revisiting it at some point in the near future.

Why Complete Streets Are Important

Bottom line: Bicyclists and pedestrians are dying! A full 13% of traffic deaths in the U.S. are bicyclists and pedestrians yet most roadways are still being built with only cars and trucks in mind.[1]

---

1.   Fatality Analysis Reporting System

Most people can see that the majority of our streets are incomplete. What most either don't recognize or don't want to upset, is the false concept that the primary use of our public streets is to move more cars and trucks faster. This concept originated in the 1950s with the push for our interstate highway system. Since then it has pushed right down to the community level severing neighborhoods and destroying historic downtowns with its blind stampede to provide speedways.

The concepts of complete streets and complete streets policies stand in the way of this blind stamped and help shift everyone's views about public rights-of-way and what our transportation networks are for. Complete streets concepts show clearly that our street systems are not just for moving more cars and trucks faster, but are the meeting spaces of our communities, for all to use.

This?

Or this?

courtesy of Michael Ronkin

While complete streets are sought by bicyclist and pedestrian advocates, advocates for transit users and people with disabilities are also vital partners. Complete streets are for everyone, and you, via your campaigns and implementation, will be able to help others and possibly broaden your constituency.

Gaining a complete streets policy helps institutionalize multi-modal transportation planning. This furthers your bike/ped-friendlier-community goals by making sure that transit options are available and accessible to bicyclists, pedestrians and people with disabilities. And, as with bicycle and pedestrian street provisions, it is also less costly for you since you don't have to dig-in and fight on every transit improvement.

How to Use this Model Campaign Document

You, as a leader of a Thunderhead organization, are one of the most important components in a complete streets campaign. You probably already have the ability, and since you are taking the time to read this document, you will soon have some of the additional knowledge that could help you be successful.

The document is structured to:

- give you a background on complete streets,
- present you with detailed information from existing and planned policies,
- suggest to you ways to help implement a policy,
- provide you with a framework of how to play out a successful campaign,
- offer you a toolkit of complete streets communication materials.

You can read it straight through, but you will also find plenty of stand-alone material in each chapter. Chapter 2 builds from the Thunderhead Alliance <u>Complete Streets Report</u> originally published in December of 2004 and provides all that we know so far about existing complete streets policies. It includes recommendations about specific aspects of policy development and strategic direction gleaned from past practice and it ends with summary recommendations on policy language. Chapter 3 gives you a taste of some of the challenges of actually implementing a complete streets policy—we will expand this chapter in all future updates and hope to include your experiences. Chapter 4 provides you with advice on starting and carrying out a campaign. Don't miss the seven elements of campaign planning, with three advocacy campaigns used as examples in each. Chapter 5 is a communications toolkit with several sections that can stand alone in guiding communications and coalition-building activities in your campaign. And don't ignore the appendices; they include a worksheet and lots of valuable information that is referred to throughout the document.

Each advocacy organization brings to complete streets a set of experiences and expertise that is unique. Each agency or jurisdiction targeted for complete streets policies is also unique in so many ways. While the combinations and variations of actions and outcomes could be seen as mind boggling, please read on and your campaign will soon take shape. If it becomes clear that your organization is not ready to pursue a complete streets policy (e.g. it may not be a wise choice for an all-volunteer organization to take on a multi-year state legislative campaign in a tough political climate), don't be discouraged or put this document on the shelf. Complete streets concepts can be used to affect change in your community with as little effort as using it in your everyday communications. What you will learn here on campaigns can also be transferred to just about any project/program you and your organization decide to tackle.

We hope this Guide provides you with the tools you need to succeed. And as mentioned earlier, your experience will be invaluable to our updates of this Guide so that more Thunderhead organizations can win complete streets policies for their communities. Your unique innovations and adaptations of these tools will rupture our current complete streets expectations and take these campaigns to new heights. But in order for this to happen, we need to hear from you. Please email your campaign and implementation experiences as well as suggestions to Sue Knaup, Thunderhead's Executive Director, at <u>sue@thunderheadalliance.org</u> so that we can weave them into our next update. We will also add you to our list of champions who have made Thunderhead's National Complete the Streets Campaign possible.

2

Complete Streets Policies

Introduction

Complete streets policies represent a potentially powerful tool for you and your organization. They are the next step in transforming your streetscapes and your communities.

As the national coalition of state and local bicycle and pedestrian advocacy organizations, the Thunderhead Alliance invested in a national survey and analysis of complete streets policy statements, directives, legislation, resolutions, plans, ordinances, and design manuals that require routinely building and reconstructing streets to be safe and convenient for all users, including those on foot and bicycle. This chapter summarizes the results of the inventory of jurisdictions with some form of complete streets policy and adds information about policies we learned about or which were adopted since the survey was completed in December 2004. It makes specific recommendations for creating effective complete streets policies and campaigns.

Methodology

This analysis of complete streets policies was derived from a survey sent to leaders of Thunderhead organizations and state and local bicycle-pedestrian coordinators throughout the United States (see Appendix C for a copy of the survey form), as well as information informally collected on new and newly discovered policies. Respondents were self-selected, although an extra effort was made to get responses from jurisdictions where policies were known to be in place The fact that the responses came from both agency staff and Thunderhead leaders means that, in some cases, different perspectives are reflected for a single policy. The two-part survey concentrated on the characteristics of the policy and on the steps taken that led to its adoption.

The baseline criteria for inclusion of a policy discussed in this chapter included: 1. calling for routine accommodation of walking and bicycling as a requirement, not as an option, and 2. covering all roads under the jurisdictions' control (this excludes bike/ped plans that only call for accommodation on certain streets). There was no evaluation on the effectiveness of these policies on the ground. However, since the survey came out, the National Complete Streets Coalition, a collaborative effort of organizations working for complete streets including the Thunderhead Alliance, has developed a standard for effective complete streets policies posted at: www.completestreets.org. For a list of active Coalition organizations see Chapter 5. Also, the Thunderhead Alliance has developed a Complete Streets Policy Checklist based on these recommended elements (see Appendix F) to help with evaluation of future policies. We have also become more familiar with what really works to create complete streets.

It should also be recognized that there is no perfect complete streets policy. Jurisdictions have taken a variety of different approaches, so these policies defy easy characterization. In addition, a policy that looks good on paper may have been essentially ignored within an agency, while a seemingly weak policy may have been implemented with gusto by local planners. So we define a good complete streets policy as one that achieves a planning, design and project development process with a constellation of new training, new procedures and design manual changes that put bicycling, walking, and transit on a par with motor vehicles. This chapter is the beginning of a learning curve, not a definitive account.

The Complete Streets Policy Checklist (Appendix F) still does not measure which policies are resulting in good outcomes on our roadways and in our communities. This will be an essential step for the future including performance measures. In addition, the analysis stops short of delving into the many design issues concerning completing the streets.

Although we made our best attempt to assemble all existing complete streets policies, there are likely some that were missed. If your community or state has a policy in place fitting the description in this chapter, please email info@thunderheadalliance.org. Thank you for your assistance.

The Policies (model and "paper" policies)

Where are policies being adopted, and what form do they take? We have learned that most existing policies are at the state level (14 total). Eleven were adopted by cities, Eight by MPOs, and three by counties. A chart of these policies, their form, and where they were adopted is shown below. A breakout of model and paper policies comes later in this chapter. And a detailed description of these policies surveyed is provided in Appendix D.

POLICIES COLLECTED	State	County	MPO/RDC	City
Legislation	OR, FL, RI, MA, MD			
Ordinance				Columbia-MO
Resolution	NC, SC	DuPage-IL	Columbus-OH	Sacramento-CA
Tax Ordinance		Sacramento-CA San Diego-CA		
Internal Policy	TN, CA, KY, VA,		Cleveland-OH Bay Area-CA	Charlotte—NC
Plans	VT		Knoxville-TN St. Joseph-MO, St. Petersburg-FL FL-AL Austin-TX	Boulder-CO Santa Barbara-CA San Francisco Colorado Springs Fort Collins-CO West Palm Beach St. Louis
Design Manual	PA, MO			San Diego-CA

<u>What does the Federal Guidance policy say?</u> Because a number of the state and local policies are based on statements in the USDOT Design Guidance, a review of that document is pertinent here (see Appendix E, Example 1 for the full Guidance text). While the language in TEA-21, where it originated, fell short of requiring states to accommodate bicyclists and pedestrians, the subsequent Guidance recommends that each state make such accommodation routine. The guidance states that:

> *...bicycle and pedestrian ways shall be established in new construction and reconstruction projects in all urbanized areas unless one or more of three conditions are met.*

The USDOT Design Guidance also calls for paved shoulders on rural roads and designs that are accessible for disabled people. It recommends using the best currently available design standards and guidelines. In a more general discussion of the approach to implementation, it recommends re-writing design manuals to include safe bicycle and pedestrian facilities while applying engineering judgment to roadway design.

The USDOT Design Guidance lists additional steps that should be taken, including:

- planning for the long-term anticipating future bicycle or pedestrian use,

- addressing the need to cross roadways, and

- requiring that exceptions be approved at a senior level and documented with supporting data.

With regard to exceptions, the Guidance lists three. They are where:

- the costs are excessive (defined as more than 20% of project costs),

- there is an absence of need (including future need), and

- bicyclists or pedestrians are prohibited from traveling by law.

The Thunderhead Alliance has developed a list of ways to enhance this Guidance for use in developing new complete streets policies. See these recommendations later in this chapter.

We use the term 'policies' loosely, because they take many forms. At the state level, five states have passed legislation: Oregon, Florida, Rhode Island, Massachusetts, and Maryland. Two states have policies that were issued by their State Transportation Commissions (North and South Carolina). Most other states have DOTs that have issued internal policies or directives.

The policies at the city, MPO, and county level include city and MPO plans, local resolutions and ordinances, and local design manuals. Some of the newest policies are tax ordinances in San Diego and Sacramento, California (approved by voters in November 2004).

Another way to analyze the policies is to look at the split between those achieved primarily through public or inherently political processes (interaction with elected officials or other political bodies) and those achieved through internal agency processes. Of the 36 policies, 13 are laws, resolutions, or ordinances and 23 are internal policies, plans, or design manuals. In several cases the internal agency-driven processes were greatly influenced by outside agents, particularly bicycle and/or pedestrian advisory groups. These policies may have also had to go through a public approval process. In addition, a comprehensive complete streets policy may take shape at several levels: first as a general policy statement in a resolution passed by an elective body, then fleshed out with administrative policies set by the implementing agency.

It is encouraging to see that complete streets policies can be achieved in many different ways at different government levels. While the statewide policies would be expected to have the most widespread effect, they commonly affect only state-owned and state-maintained roads. Oregon's state law is an exception as it affects all roads, no matter the jurisdiction. Other state polices may influence local communities and lead to the creation of more local policies. In California for example, Deputy Directive 64 seems to have spurred additional local action.

We have also discovered some complete streets policies that we call 'paper policies' because they look good on paper but are not being implemented. Bringing these policies to light is important in helping Thunderhead leaders and agency officials begin to work on their full implementation. See the implementation chapter for more details.

In the more detailed table below, you will find paper policies listed below model policies. The model policies are highlighted due to the fact that the leaders of the Thunderhead organizations serving those areas have found them to be helpful to their bicycle and pedestrian advocacy efforts. The paper policies have not yet been helpful to the Thunderhead leaders.

When were policies adopted? The move toward complete streets has been growing. Most have come about since 2001, and a significant portion were adopted in 2004 and 2005. This is in part a testament to the influence of the 2000 USDOT Design Guidance, "Accommodating Bicycle and Pedestrian Travel," which was issued in response to language included in the Transportation Equity Act for the 21st Century (TEA-21). This Guidance is an important base for many complete streets policies. A few of the inventoried policies precede this era. For example, Oregon's was enacted in 1971 and offers an opportunity to evaluate longer-term impacts of these policies.

What do the state and local policies say? It is important to note that of all the policies included in the survey, only a few of the policies, laws, resolutions, ordinances, plans, or design manuals use the term 'complete streets.' Nonetheless most of these policies have great language setting out their vision. A few examples follow.

> *…bicycling and walking accommodations should be a routine part of the Department's planning, design, construction and operating activities.*
> (SC Department of Transportation Commission resolution)

> *Bicycle and pedestrian ways shall be established in new construction and reconstruction of road and bridge projects unless one or more of four conditions are met.*
> (Cleveland, Ohio MPO)

> *Footpaths and bicycle trails {bikeways and walkways} including curb cuts or ramps as part of the project, shall be provided wherever a highway, road or street is being constructed, reconstructed or relocated.*
> (Oregon statute)

> *This document outlines an approach to designing streets that are more "complete" in the sense of accomplishing all of the goals associated with the dominant form of public space in urban societies—our streets.…Complete streets are those that adequately provide for all roadway users, including bicyclists, pedestrians, transit riders, and motorists, to the extent appropriate to the function and context of the street.*
> (Sacramento, CA Best Practices for Complete Streets)

Policy Issues

Does the policy really *require* accommodation? Many jurisdictions have plans and policies that express a *desire* to ensure the road serves all users. The most basic element of any complete streets policy is that it ensures that roads are built with everyone in mind. In some cases, policies use the word "consider." For example,

The Department fully considers the needs of non-motorized travelers (including pedestrians, bicyclists and persons with disabilities) in all programming, planning, maintenance, construction, operations and project development activities and products.
(CalTrans Deputy Directive 64)

This should raise a red flag for Thunderhead leaders, because 'consideration,' in the words of one Thunderhead leader, can give agencies "tons of wiggle room." That said, the California policy has been used effectively by Thunderhead leaders to press for localized complete streets initiatives. The way to turn 'consideration' into a more robust policy is to establish clear guidelines for what it means: filling out a checklist, getting approval of exceptions, etc. Better yet, avoid the terms "consider" and "consideration" choosing instead stronger language such as "shall be included in every project."

And always be sure to read beyond the initial lofty statement. Even with strong language in the initial statement, some policies may not function as complete streets policies. For example, while Arizona has a policy which states "It is Arizona DOT's policy to include provisions for bicycle travel in all new major construction and major reconstruction projects on the state highway system," the many exceptions and restrictions that are listed just after this statement set up hurdles that make it clear that providing complete streets will occur only in special circumstances, not as a matter of course.

Our RECOMMENDATION is that you use stronger "shall be established" or "shall be included" language instead of "consider." These will, in effect, require accommodation to be a routine part of all road design and re-design.

Exceptions: A more precise way to get at whether policies truly require complete streets is by looking at any specific exceptions, and how those exceptions are handled. By setting a rigorous, formal process for approving exceptions, agencies create a process that helps ensure compliance. Some of the policies list specific exceptions, including:

- excessive cost,
- absence of need,
- lack of right of way, and
- no need during simple repaving projects.

Other exceptions specified in some policies are public safety, environmental considerations, project purpose and scope, low traffic volumes, and conflicts with local plans. These exceptions go far beyond the USDOT Design Guidance, which lists three limited exceptions. As discussed previously in this chapter, these are:

- excessive cost,
- absence of need, and

- where bicyclists and pedestrians are prohibited.

The USDOT Guidance defines excessive cost as more than 20% of project costs and specifies that need should be defined in terms of potential *future* pedestrian or bicycle travel (we all know about the potential for significant latent demand).

Remember the Cost Misconception: A common misconception is that complete streets cost more to build than incomplete streets. In fact, complete streets most often cost no more and many times can cost less than incomplete streets. For instance, a common street cross section that serves only cars is a four lane speedway with no shoulders, sidewalks or intersection treatments for people. Using the same right-of-way width, this design can be reshaped into two narrower through lanes, one center turn lane, and bike lanes and sidewalks on both sides. By using less width for the most expense elements, truck weight standard asphalt and subsurface, and adding less expensive sidewalks, this design, often referred to as a "road diet" when applied to existing roads, actually saves money. Not only that, this design has been proven to improve traffic flow and safety for motor vehicles by better controlling turning movements. Many other complete streets designs offer similar cost savings. You may even want to bring up the economic benefits of streets that attract visitors and offer access to more employees. Be sure to address this misconception early in your campaign so that you can focus your valuable time on instituting a policy for your communities.

When America Bikes, the coalition of eight national bicycle advocacy organizations working on the reauthorization of TEA-21, the federal transportation law, was seeking to place complete streets language in the new law, costs seemed to be a primary issue with members of Congress. America Bikes collected statements from DOT officials who said that integrating bicycle and pedestrian provisions from the beginning should not significantly increase costs. Of course one of the beauties of a complete streets policy should be that bicycle and pedestrian facilities are no longer fighting for the small pie of funds specifically designated for bicycling and walking (such as Enhancements or CMAQ), but are simply part of general transportation spending.

In line with these statements, cost did not seem to be a primary implementation issue for survey respondents. A few respondents did note that once initial budgets are set, including bicycle or pedestrian provisions can become almost impossible. Others noted that right-of-way acquisition can be the most expensive part of a road project, so wider roads with bike lanes may be a barrier. In such cases, reducing the number of travel lanes, otherwise known as a road diet as mentioned above, can complete the street actually at a cost savings.

It should be noted that the most common exception allowed is 'excessive cost,' often set at 20 percent of project cost. Michael Ronkin said it is important to be specific about what constitutes 'total project cost' since many projects are broken down into smaller parts. Sidewalks may be a significant cost if the project is defined as paving of a one-mile road subsection, but may make up a smaller portion when the project is defined more broadly to include all improvements in the whole corridor.

> Our RECOMMENDATION to you is that if your policy includes an "excessive cost" exception, make sure that it clearly states the broadest scope of the project so that sub-section cost breakouts are not possible.

<u>Exceptions Approval Process:</u> The next question is whether the policies require any formal approval when exceptions are made and all modes are **not** accommodated. The USDOT Guidance recommends that such exceptions should include documentation and require approval from senior management. Just nine of the 36 policies require such formal justification. The survey form did not ask about the exact method for documenting justifications, but in some cases survey respondents mentioned that there are design exemption forms or required checklists. Thunderhead leaders noted that a formal exemption process was valuable. One leader put it this way:

> *At least now, the engineers have to file a formal 'design exemption' outlining the reasons for not including bike or ped accommodation instead of just not doing it.*

> Our RECOMMENDATION is that you should work for policies that have a limited set of exceptions, if any, and that require a formal approval process for each exception. Policies should reverse the current norm from having to justify accommodating all modes to having to justify NOT accommodating them.
>
> While a reluctant agency can still find ways to use exemptions and other language to exclude accommodation, the process gives Thunderhead leaders both leverage and the opportunity to work with and change the attitudes of reluctant engineers and planners. At the end of this chapter, there are further recommendations for crafting policy language, as well as examples of good language already in use.

<u>Design specifications:</u> Another issue is how prescriptive the policies are with regards to actual street design. Few of the policies provide specific language on what types of accommodation should be undertaken (e.g. when and where to build bike lanes or add sidewalks with curb-and-gutter, etc) unless the policy is itself a design manual. Most of the documents are, instead, broad policy statements that refer to other guidelines or design manuals for design specifics. In some cases, jurisdictions have achieved complete streets by revising their standard street cross-sections to include other modes. The USDOT Guidance recommends that agencies should "design facilities to the best currently available standards and guidelines," mentioning AASHTO and ITE standards.

> Our RECOMMENDATION is that you steer away from specifying design standards in your policy, especially in an initial complete streets policy campaign. The discussion of the intent (a commitment to build streets for all users) should be separated from the design discussion. As Thunderhead leaders, your role is to push for the *vision* of complete streets. Getting bogged down in arguing about narrow specifications could be deadly to the overall effort.

<u>What modes do the policies cover?</u> The ideal complete streets policy makes clear that roads must be built and reconstructed to serve all users including pedestrians, bicyclists, transit users, and travelers of all ages and abilities. Few of the existing 36 policies are that comprehensive. Several of the policies discuss accommodating transit and people with disabilities, but many do not. The USDOT Design Guidance makes specific reference to accommodating people with disabilities as follows:

> *The 1990 Americans with Disabilities Act, building on an earlier law requiring curb ramps in new, altered, and existing sidewalks, added impetus to improving conditions for sidewalk users. People with disabilities rely on the pedestrian and transit infrastructure, and the links between them, for access and mobility. (USDOT guidance)*

A few notable examples incorporate transit elements. For example, see San Francisco's Transit First policy. The Sacramento Transportation and Air Quality Collaborative's "Best Practices for Complete Streets," includes a section on designing the road for transit users, noting that, *"The key design issue in planning for transit is the out-of-vehicle time (time spent waiting and time spent walking to and from the transit stop) which often plays a more important role in the decision to use transit than time spent in the vehicle itself."* Essentially, planning for transit is planning for pedestrians, and even for bicycle users, as bike-on-bus programs continue to expand.

<u>How do bicycle and pedestrian plans fit in to complete streets?</u>
Complete streets policies are about integrating all modes of travel into a single design process. Many communities have adopted stand-alone bicycle and pedestrian plans and design manuals which have helped created much of the progress we've seen in the last 20 years. However, these plans have often failed to result in true integration, and can even foster competition among modes. This was the case in Boulder, Colorado, which discovered that an integrated approach ended in-fighting between transit, bicycling, and pedestrian programs. Also, plans often only list specific streets for accommodation rather than all streets as with complete streets policies.

Our RECOMMENDATION is that you seek complete street policies that incorporate transit and active living. Why? This is one of the most significant differences between 'routine accommodation' and 'complete streets.' If complete streets by definition provide safe travel for all users, and if part of the intent of pursuing complete streets is to build alliances beyond bicycle and pedestrian concerns, advocacy leaders seeking to build alliances in a broad complete streets campaign will need to amend the language to discuss other issues.

The US DOT Design Guidance advocates this approach. In a section called "Rewrite the Manuals" specific bicycle/pedestrian manuals are portrayed as an interim step toward a recommended total rewrite of general street design manuals. At the same time, the

Guidance also recommends allowing 'engineering judgment' to guide decisions on a case-by-case basis. All of the examples given show circumstances in which *more* bike/ped accommodations should be made than those identified by design standards.

> Our RECOMMENDATION is that you follow Oregon's example, if possible, and keep your policy language non-specific to responsible agencies.

What roads are covered? Most of the 36 policies cover only those roads that are under the direct responsibility of the agency in question. For example, many of the state DOT policies only cover state-owned roads. In the case of MPOs, they tend to cover roadway projects funded through MPO-disbursed funds (which are usually federal transportation dollars). The new sales tax ordinances in Sacramento and San Diego counties apply to all the projects funded under the ordinances. A few of the local policies are directed at developers building new subdivisions. Michael Ronkin, Oregon DOT Bicycle and Pedestrian Program Manager, notes that the passive grammar of Oregon's state law has helped ensure that it applies to every road. Oregon's law says, "wherever a road is constructed" without referring to the agency responsible for building or maintaining it.

Funding: Most of the policies identified do not include specific funding provisions. The USDOT Design Guidance does not mention funding (except a suggested restriction on excessive cost). The notable exception is Oregon, which set aside one percent of its state transportation funds for bicycling and walking facilities. More often, the policies make bicycle and pedestrian accommodation a prerequisite for funding that already exists—the MPO policies and the tax ordinances specify that funded projects must accommodate travel by alternative modes, usually foot and bicycle. The other policies usually assume that funding will come from standard sources. But, again, remember the misconception that complete streets always cost more. See more about this misconception earlier in this chapter.

> Our RECOMMENDATION is that you think through funding issues ahead of time and identify, if possible, a funding stream for the policy for those complete streets projects that will add costs. This, along with a strong message that complete streets often do not cost more than incomplete streets, will help you secure your policy.

One Thunderhead leader mentioned that their state's restriction on spending gas-tax money only on roads may get in the way of local jurisdictions' implementation on their new MPO policy. Thirty states have such a restriction on the books, but it is unclear whether they have actually prevented funding of bicycle and pedestrian projects.[1]

So, what is a good policy?

All of this discussion makes complete streets policies seem pretty complex. To simplify things, we tried to distill the elements that do the most to contribute to that change in agency culture that leads

---

1.   A list of state restrictions can be found in the Brookings Institution report, *Fueling Transportation Finance: A Primer on the Gas Tax* http://www.brookings.edu/es/urban/publications/gastax.htm.

to full integration of all modes. They include: inclusion of as many modes as possible, a process that requires any exceptions to be approved at a higher level, and a clear definition of those exceptions. We also checked on what implementation steps have been undertaken, and whether Thunderhead leaders deem the policy useful (even if it is not perfect). The table below gives the results of this scan highlighting those policies that have been helpful as models. You will find these model policies marked on the map for Thunderhead's National Complete the Streets Campaign at: http://www.thunderheadalliance.org/completestreets.htm . Our goal for this campaign is to help our organizations win at least one model complete streets policy, local or state-level, in all 50 states by 2008 in order to influence a model federal-level complete streets policy through the reauthorization of SAFETEA-LU.

Complete Streets Policies Table

Model policies followed by "paper" policies

State	Project title	Users?	Senior-level approval required for exceptions?	Extra Exceptions allowed (beyond cost, no need, prohibited)	Implementati on steps undertaken	Thunderhead org leaders have found policy helpful
Model Policies						
CA	California Dept of Transportation Deputy Directive 64 internal policy	ped, bike, disabled	no	exceptions not specified	updated pro-cedures; more?	**yes**
CA	Sacramento rou-tine accommoda-tion sales tax initiative	bike, ped	no	none speci-fied	unknown	**yes**
CA	Sacramento	bike, ped				**yes**
CA	Bay Area MPO (MTC) Second Cycle Program-ming Policies, screening criteria	bike, ped	no	exceptions not specified	unknown	**yes**
CA	Santa Barbara Cir-culation Element, General Plan	all	no	insufficient ROW do not plan separate bike facilities on roads with 25 mph limits	unknown	**yes**

State	Project title	Users?	Senior-level approval required for exceptions?	Extra Exceptions allowed (beyond cost, no need, prohibited)	Implementation steps undertaken	Thunderhead org leaders have found policy helpful
Model Policies						
CA	San Diego City Street Design Manual		yes	Excessive cost Insufficient ROW	re-written manual	**yes**
CO	Colorado Springs Complete Streets Amendement to the Intermodal Transp. Plan	ped, bike, transit	not stated	unsafe impractical	rewriting manuals	**yes**
CO	Ft. Collins Colorado	ped, bike, transit	yes	none	restructured procedures—(LOS) rewritten design manuals	**yes**
CO	Boulder Multimodal Corridors & Transportation Network Plans	ped, bike, transit	yes	none	restructured procedures re-written manuals training	**yes**
FL	West Palm Beach FL Transportation Element	ped, bike	not stated	exceptions not specified		**yes**
FL	Florida Bicycle & Pedestrian Ways statute	ped, bike	yes	excessive cost absence of need where contrary to public safety	unknown	**yes**
IL	DuPage County Healthy Roads Initiative	ped, bike	not stated	exceptions not specified	unknown	**yes**

State	Project title	Users?	Senior-level approval required for exceptions?	Extra Exceptions allowed (beyond cost, no need, prohibited)	Implementation steps undertaken	Thunderhead org leaders have found policy helpful
Model Policies						
MA	Bicycle-Pedestrian Access Law, Massachusetts staet legislature (Chapter 90E)	ped, bike	Yes	discretion of commissioner, safety environmental quality ROW conflicts	unknown	**yes**
MO	St. Louis Legacy 2030 Long-Range Plan	ped, bike, transit	not stated	no exceptions specified	checklist	**yes**
MO	Columbia Missouri Model Street Standards	ped, bike	No			**yes**
MO	St. Joseph MO bike-ped plan	ped, bike	yes	shoulders on rural roads	unknown	**yes**
NC	North Carolina DOT Bicycle Policy	ped, bike	No		unknown	**yes**
OH	Columbus Ohio MPO (MORPC) Bicycle and Pedestrian Planning Policy	ped, bike	yes		unknown	**yes**
OH	Cleveland Ohio MPO (NOACA) Regional Transportation Investment Policy	ped, bike	yes	extreme topography/ natural resource constraints low ADT—below 1,000 simple resurfacing projects	unknown	**yes**
OR	Oregon Bicycle and Pedestrian Statutes	ped, bike	yes	public safety	restructured procedures re-written manuals training	**yes**

State	Project title	Users?	Senior-level approval required for exceptions?	Extra Exceptions allowed (beyond cost, no need, prohibited)	Implementation steps undertaken	Thunderhead org leaders have found policy helpful
Model Policies						
SC	South Carolina DOT Commission Resolution	ped, bike	not stated	exceptions not specified	restructured procedures training	**yes**
TN	Tennessee DOT Bicycle and Pedestrian policy	ped, bike	yes	bridges insufficient ROW repaving	unknown	**yes**
VA	VDOT Policy for Integrating Bicycle and Pedestrian Accommodations	ped, bike	Yes	environmental impacts safety purpose & scope of Project	none	**yes**
VT	Vermont Bicycle Pedestrian Plan	ped, bike	not stated	not specified	training	**yes**
Paper Policies						
CA	SF Transit First policy city ordinance	ped, bike, transit	not stated	not specified	unknown	no
CA	San Diego County Transnet Tax Extension provision	ped, bike	not stated	exceptions not specified	unknown	too early to say
FL-AL	Florida-Alabama Transportation Planning Organization (TPO (bicycle plan)	ped, bike	not stated	no exceptions specified	updating procedures	too soon to tell
FL	St. Petersburg "citytrails" plan					no
KY	Kentucky Pedestrian and Bicycle Travel Policy	ped, bike	not stated	exceptions not specified	none	no
MD	Maryland Transportation Code Ann. 2-602	ped, bike	not stated	exceptions not specified	none	unknown

State	Project title	Users?	Senior-level approval required for exceptions?	Extra Exceptions allowed (beyond cost, no need, prohibited)	Implementation steps undertaken	Thunderhead org leaders have found policy helpful
Paper Policies						
MO	p. 24-25 of MoDOT's Practical Design Implementation Manual	ped, bike				Not yet
NC	Charlotte Urban Street Design Guidelines internal policy	ped, bike, transit	yes	None	restructured procedures	not yet
PA	Penn Bicycle & Ped Checklist Training (App. J to PennDOT Design Manual)	ped, bike	no	exceptions not specified	checklist	no
RI	Rhode Island state law and policy	ped, bike	no	public safety, environmental or scenic quality, ROW conflict at Director's discretion	unknown	no
TN	Knoxville MPO Bicycle Accomm. Policy		yes		unknown	not yet
TX	Capital Area MPO, Texas Mobility Plan 2030	ped, bike	not stated	demonstrated alternative plan	unknown	unknown

Overall Recommendations for Policy Development

First, here are some concluding policy observations:

1. Policies take many forms and have been adopted at all levels of government, with adoption accelerating in recent years.

2. Policies vary in how strict they are in requiring accommodation. Some have set specific exceptions. Most policies do not themselves give design specifications. Despite imperfections, Thunderhead leaders see policies as providing important leverage for their efforts.

3. Most policies focus almost exclusively on bicycling and/or walking and do not significantly discuss transit users, people with disabilities, or other user groups.

4. Implementation issues are significant; the work does not end with policy adoption.

5. No policies include effective performance measures, and little data is being collected on how well they are working.

Also, we recommend including these elements specified in the "Elements of Complete Streets Policies" on the complete streets web site: www.completestreets.org :

ELEMENTS OF COMPLETE STREETS POLICIES

1. The Principle

- Complete streets are designed and operated to enable safe access for all users. Pedestrians, bicyclists, motorists and transit riders of all ages and abilities must be able to safely move along and across a complete street.

- Creating complete streets means changing the policies and practices of transportation agencies.

- A complete streets policy ensures that the entire right of way is routinely designed and operated to enable safe access for all users.

- Transportation agencies must ensure that all road projects result in a complete street appropriate to local context and needs.

2. Elements of a Good Complete Streets Policy
A good complete streets policy:

- Specifies that 'all users' includes pedestrians, bicyclists, transit vehicles and users, and motorists, of all ages and abilities.

- Aims to create a comprehensive, integrated, connected network.

- Recognizes the need for flexibility: that all streets are different and user needs will be balanced.

- Is adoptable by all agencies to cover all roads.

- Applies to both new and retrofit projects, including design, planning, maintenance, and operations, for the entire right of way.

- Makes any exceptions specific and sets a clear procedure that requires high-level approval of exceptions.

- Directs the use of the latest and best design standards.

- Directs that complete streets solutions fit in with context of the community.

- Establishes performance standards with measurable outcomes.

2.5 Implementation
An effective complete streets policy should prompt transportation agencies to:

- Restructure their procedures to accommodate all users on every project.

- Re-write their design manuals to encompass the safety of all users.

- Re-train planners and engineers in balancing the needs of diverse users.

- Create new data collection procedures to track how well the streets are serving all users.

Sample Policies

Many Thunderhead leaders and agencies have asked for sample complete streets policy language. Such samples are difficult to craft, as every jurisdiction has unique needs. A solid complete streets policy should:

a. require accommodation as a routine part of all road design,

b. set a clear procedure for specific exceptions that requires formal, high-level approval, and

c. direct agencies to use the best available design standards and guidelines.

For more details, see "Elements of a Complete Streets Policy" (above and on the complete streets web site). Links to a variety of existing policies can be found in the appendices of this Guide and on the complete streets website; finding a policy close by can be an effective starting point. Also see the Complete Streets Policy Checklist (Appendix F).

Starting with the US DOT Design Guidance

Since 2000, most of the strong complete streets policies have been modeled after the USDOT Design Guidance: Accommodating Bicycle and Pedestrian Travel (see Appendix E, Example 1) which includes a solid policy statement that can, and has been, adapted for a number of different formats and holds credibility with transportation agencies. Here are some ways it can be improved upon.

- Add a compelling case statement at the top. See Appendix E, Example 2, the introductory text to the MORPC Bicycle and Pedestrian Planning Policy. We suggest using the phrase 'complete streets' instead of 'routine accommodation.'

- Make sure you use stronger "shall be established" or "shall be included" language. Do not allow your agency, as some have done, to borrow the weaker points and very weak "consider" language from TEA-21.

- Look at eliminating a specific percentage for excessive cost, or specify that the percentage covers the entire project, as opposed to a single road segment. The 20 percent, oft-used figure for excessive cost has been disputed in some cases.

- Elevate two important points that are somewhat buried in item 4 of the USDOT Design Guidance:

  - that 'scarcity of need' should be considered in terms of future, rather than current use, and

  - that exceptions should be approved at 'a senior level' and build on this by requiring the agency to justify not accommodating bicyclists and pedestrians through a detailed process.

- Add language to clarify the need to accommodate transit vehicles, transit users, as well as people with disabilities. To date, only a few policies include transit, and none follow the format of the Design Guidance.

- Consider adding language on measurement of progress toward creating complete streets.

Thunderhead leaders who are looking for a more general resolution on complete streets may want to consider the South Carolina Department of Transportation Commission's resolution (Appendix E, Example 3).

The Policy Adoption Process

This section shares the experience of some engaged in campaigns to get complete streets policies adopted in their communities, as well as some information about how some of the existing policies came to be. This is a supplement to the step-by-step campaign planning information in Chapter 4; don't skip those important steps!

At least twenty state and local complete streets campaigns were underway through Thunderhead organizations at the time of this writing. Three of these offer interesting insight into the process. While none had yet met success, the Thunderhead leaders engaged in them have already learned much about what in complete streets resonates with policy makers—and what concerns them.

League of Illinois Bicyclists Statewide Campaign
The League of Illinois Bicyclists has been working for several years for a complete streets policy, and in the winter of 2006 appeared close to convincing the Governor to adopt the policy through administrative means. Ed Barsotti, Executive Director of LIB, recommends first going to the agency in question and asking for the policy. This helps build a relationship with the agency that ultimately has to implement the changes—even if they say no at first.

After IDOT was not responsive, LIB worked with state legislators to submit a bill based on the federal design guidance. LIB asked for—and received—support from two statewide disabled advocacy groups, the Illinois Public Health Association, and the Illinois PTA. LIB promised to do the legwork, but they expect to expand their work with these groups in the future. In the State Senate, LIB had to cope with a DOT analysis that overestimated the cost of implementation. Despite this the bill passed with a comfortable margin in the spring of 2005. However, it became hung up in the House due to factors unrelated to its content.

The next step was to enlist the help of the Governor's office. After months of trying, a number of advocates were able to meet with one of the Governor's staffers. At this meeting, Ed and others presented the case for complete streets, with a heavy emphasis on safety—a strong interest of the Governor.

The presentation relied heavily on photographs of recently-constructed projects with no bicycle or pedestrian facilities. They told the story of a bridge in Cary, Illinois, which crosses the Fox River. The bridge has no space for non-motorists, and within three years, three teenagers died trying to cross the river by foot or bicycle: one bicyclists using the road's median, one teen crossing a railroad trestle who was hit by a train, and one teen who drown trying to cross the river. These stories proved very effective in getting the staffer on their side. You can see the presentation at http://www.bikelib.org/completestreets/sb508mtg1005.pdf.

"I wish we'd done a more visual presentation from the start, when we first met with DOT officials," says Ed. "At the early DOT meetings we went through policies and specified needed changes. But it seems more effective to go out and find recently built, inadequate projects, take pictures, and then specify the policies that allowed them to be built that way."

Bicycle Alliance of Washington local and statewide campaigns
The Bicycle Alliance of Washington has been pursuing complete streets policies at both the state and local level. For the City of Seattle, the Alliance countered initial resistance by putting together a 'design collaborative' to document and then discuss the deficiencies in the bike network. Teams of volunteer cyclists were assigned to go out and document conditions. Their findings were shared at a special design collaborative meeting with City Council members, which was televised. The Alliance's efforts created a process and put important information in front of council members. While the final policy is stalled because of city politics, Executive Director Barbara Culp believe success is coming. She credits the collaborative approach between city officials, city transportation professionals, and bicycle advocates.

At the state level, the Bike Alliance is pursuing an internal policy, and has been working with potential allies at all levels—elected officials, agency decision makers, bicycle club members, and advocates. Culp says they have encountered little opposition, but persistence has been important. "Acknowledge that it's going to take twice as long as you imagined."

Bicycle Colorado statewide campaign
Bicycle Colorado has been talking about complete streets in every meeting, fundraiser, and email to show allies and potential allies they have a vision, a solution, and a plan to make a statewide complete streets policy a reality. They have had the idea brought up in committee meetings of the state legislature as a way to explore the best mechanism for getting it enacted. But before they begin to push hard, they are working to build a coalition of advocates for transportation alternatives, seniors, public health, and other community groups so they can present a broad grassroots campaign. "Be

prepared for a multi-year campaign," advises Executive Director Dan Grunig. "Complete streets is an idea that takes a little while to comprehend and buy into.'

How existing policies came to be

The survey of complete streets policies included questions about how the policies came about. While the questions on the survey did not ask for a full history of the effort, in five cases, bicycle and pedestrian advocacy organizations were credited with making the original push for the policies. In two instances, a bicycle or pedestrian advisory committee is credited with originating the idea.

> *Bicycle advocates and legislators urged the California DOT to adopt the USDOT Policy Statement on Integrating Bicycling and Walking into Transportation Infrastructure. The DOT preferred to develop its own policy.* (California)

In many cases Thunderhead leaders worked on the policies through the official bicycle/pedestrian committee or advisory board. For the most part, the advocacy approach on these early wins was low-key, without a lot of broad public outreach. When asked about specific activities, the most common advocate activity was attending and arranging meetings with staff and officials, and to participate through advisory boards or other official bodies. Six organizations mentioned working on writing or revising the actual policy language, with a few saying this was very valuable. Media-based public outreach tactics were mentioned by only two Thunderhead leaders reporting on policies: Virginia and Columbia, Missouri. Columbia developed an impressive set of materials as well as a broad list of allies in an effort that included media and public presentations.

This more internally focused campaign style is reflected in the allies named as part of the complete streets efforts. In the survey, six Thunderhead leaders mentioned internal allies at the agency adopting the policy. Only a few mentioned groups other than usual bicycle-pedestrian allies.

Our RECOMMENDATION is that you strengthen your organization by using complete streets to build coalitions with natural allies: public health groups, smart growth groups, transit groups, children or senior advocacy groups. See Chapters 4 and 5.

Opposition: Seven respondents in the survey indicated public resistance, including landowner resistance to wider right-of-ways, worries about costs, and concerns about safety or appropriateness of accommodation. The most organized public resistance appears to be in Santa Barbara, where their circulation element, in place since 1995, has inspired a website called Cars are Basic: http://www.sil-com.com/~cab/cab.htm.

Some respondents mentioned resistance from specific groups, including from within the DOT, from the local congestion management association (which saw the move as competing for funds), and from the development industry (in those cases where the developers are responsible for providing the roads).

A few sample comments from the survey:

People from our Board and Transportation Advisory Committee, in particular, county engineers, were leery. They insist we need a map with lines on it so they know where they really have to put facilities. At this time, NOACA doesn't have such a map and the BAC met recently to consider the idea and rejected it as inconsistent with our policy. (Cleveland MPO)

Opponents have argued that Florida DOT implementation is wasteful (i.e., that bicycle lanes are underused, relative to cost) or is unsafe—many members of the public feel that cyclists are more appropriately accommodated on separated paths. (Florida)

There is a fear that bike lanes would invite children and inappropriate users to particularly busy roads. (Illinois)

The good news is that in many cases the policies are not opposed, but may be resisted by planners or engineers mainly because they are not quite sure how to go about it. In South Carolina, initial resistance softened as the engineers applied themselves to the task of figuring out *how* to make accommodation. Thunderhead leaders can address this issue early by providing agency officials with options for training; contact the Association of Pedestrian and Bicycle Professionals for more information about consultants who can provide such assistance.

> Our RECOMMENDATION is to be alert to the concerns of opponents in your early outreach efforts, and when possible find ways to directly address their concerns. See "Element 3—Gauge Your Resources" in Chapter 4 for advice on opposition.

Keys to Policy adoption success: The survey asked Thunderhead leaders to summarize the roots of successful policy adoption in three key points. A few of their answers:

1. *Supportive, sympathetic staff at MPO.*

2. *Adoption of routine accommodation at rival MPO in northeast Ohio in fall of 2003, challenging leadership position of our MPO.*

3. *Threat to federal funding for local transportation projects if they do not adopt routine accommodation policy.* (Columbus Ohio MPO)

1. *Strong grass-roots support.*

2. *Constantly positive image in the media (we never engaged in public criticism of anyone).*

3. *Working the media.* (Columbia MO)

1. *Existence of DD64 [California statewide policy].*

2. *Supportive MTC [MPO] chairman who is a friend.*

3. *MTC prides itself on being progressive.* (CA Bay Area MPO)

Recommendations on an Advocacy Approach

The ultimate aim in pursuing a complete streets campaign is to create a culture in which every street is built, modified, and maintained to be safe, comfortable, and even inviting for users of all modes. A complete streets policy will not, by itself, achieve this goal. Agencies will be resistant; individual projects will be controversial; other priorities will prevail. It won't end the road battles that motivate so much advocacy work. Instead, the policy is best viewed as a vehicle for change.

The campaign for the policy is a way to educate decision makers and the public about prioritizing our streetscapes differently. The policy itself will give Thunderhead leaders important new leverage in pursuing better accommodations, both across the jurisdiction and in individual road battles. Most importantly, the policy will provide a way to push transportation agencies toward culture change. The process of re-writing design manuals, or training transportation agency employees in implementation, should be seen as an integral part of reaching the ultimate goal.

- Thunderhead leaders must assess their organizational strength and the political conditions in which they are working as they choose whether to immediately pursue a strong policy or to work toward complete streets in stages.

- Thunderhead leaders looking for the most comprehensive policy may consider launching a campaign for a statewide law. To date, most states that have statewide legislation achieved them as part of wider reforms.

- Thunderhead leaders seeking 'lower-hanging fruit' may opt for a policy adopted through an administrative process at a friendly agency. Internal and local policies obtained through an administrative strategy have a clear record of adoption.

- Thunderhead leaders may also engage in more modest efforts to simply spread the concept of complete streets, laying the groundwork for a future policy campaign.

In a broader sense, Thunderhead leaders should also see complete streets as just one part of making communities better for bicycling and walking. Much of what encourages people to walk, bicycle and use transit are the variety of destinations within a reasonable distance. Without land-use changes, sprawl will continue to erode the ability to walk and bicycle. Complete streets are a part of this mix because they are a way to make common cause with other organizations working for healthier communities that offer residents more choices and better access.

3

Implementation

Complete Streets Implementation Issues

Once a policy has been adopted, the hard work begins: effective implementation. A few of the policies identified in this Guide are no more than 'paper polices.' They hold promise, but little or nothing has been done to implement them and integrate new practices into agency procedures. In some cases, few people even seem to know about them. See the detailed table in Chapter 2 for a list of these policies.

Your complete streets policy campaign will initially target a specific public policy decision by the legislature or the transportation agency. It is important however that throughout the campaign you keep your eye on your ultimate goal—major changes in the way all transportation decision-making is done to achieve a balanced multi-modal outcome.

For most transportation agencies, fully implementing complete streets will mean a fundamental shift in previous procedures and assumptions. Most agencies have focused on maximizing automobile throughput, and many engineers are trained primarily to achieve this goal. A shift that requires a broad assessment of the needs of all road users does not fit easily into this paradigm.

As with any bureaucracy, a transportation agency can have systemic inertia that is comprised of individual attitudes, long-standing habits and procedures, incomplete technical knowledge, and entrenched relationships. Any broad policy change at the top will travel a long road with many smaller policy and procedural changes along the way. The motivation of the leadership of the agency to implement this policy is going to make a big difference. The way the initial policy came about will also make a big difference. If a complete streets policy was forced on a recalcitrant agency, the battle for implementation will probably be long. If the legislative or policy campaign was used to get agency officials to see value in the policy, implementation will probably be easier.

In the survey, respondents identified a number of barriers to implementation. Some said agency implementers were not aware of the policies or could not agree on what they mean. Some said no steps were established to move toward implementation, including a failure to choose or create design standards. A couple of respondents noted the difficulty of increasing the width of a right-of-way, particularly in infill areas. Other implementation issues included a failure to include facilities in initial budgets, a lack of MPO input into design, and a resistance of the state DOT in working with a local jurisdiction. Some respondents in areas with a policy directed at new development noted that it is difficult to ensure that development agreements for specific projects include complete streets, since governments are often reluctant to make such requirements of developers (note that even when such requirements come in to existence, many developers will then work hard at seeking exceptions). Thunderhead leaders also mentioned a simple lack of resolve or a bias against bike lanes as implementation barriers, while some staff respondents cited resident resistance to the changes, particularly those that increased road width.

When creating your Complete the Streets campaign, consider implementation part of the campaign. Chapter 2 reviewed some of the barriers to implementing existing complete streets policies. They range from the avoidance of turning a policy document into effective procedures, to the misconceptions of costs, to standard agency resistance. Some agency implementers will claim that they are not aware of the policies or that there is no agreement on what the policies mean. In this chapter, we will focus on working with your agency to set up an effective implementation procedure.

Keep in mind that even once the policy and procedures are in place, your organization will likely find itself fighting some familiar battles over transportation projects. It might help to think of a solid complete streets policy not as the complete solution, but as an important step in your advocacy. How can you make that tool most effective?

From Policy to Procedure

An effective, well-designed complete streets policy should prompt the following internal agency changes.

courtesy of David Crites

- Restructuring procedures to favor multi-modal planning.

- Re-writing design manuals.

- Retraining planners and engineers.

- Re-tooling measures to track outcomes (there is the possibility that they may not be tracking any outcomes now).

> Our RECOMMENDATION is that you simply understand that there will be some barriers. You will need to stay involved, even help, in the initial implementation stages and then check back periodically.

Your influence over this internal process may be formal, through an advisory committee, or informal, through your relationships with agency staff. Respect the agency's process and try to position yourself as a resource. You may be able to increase the credibility of your suggestions by referring to experience at other agencies and the recommendations made in the USDOT Design Guidance.

Your ongoing relationship with the legislators and elected officials that led to the initial policy change is a key to your influence on the agency. You will build respect and influence if you are seen as the one who communicates progress, or lack of progress, back to the people that they are accountable to.

Your strong relationship with and handling of the media also impacts your influence on an agency and with legislators and elected officials.

Restructuring procedures: Some agencies will see an opportunity in a complete streets policy to take a whole new approach to transportation planning, moving away from the traditional focus on volume-to-capacity ratios and Level of Service determinations. For example, Charlotte, North Carolina, in an effort to turn their paper policy into a model is instituting a new six-step planning process that begins by establishing the land use and transportation context of the project, identifying gaps and deficiencies in the network for all users, and then engaging in a clear process to meet the challenge of balancing the needs of all users. Boulder, Colorado has also developed a planning process to conduct an initial evaluation of the needs of *all* users. Thunderhead leaders can make agencies aware of these opportunities to create fundamental change.

Other agencies will prefer to look for ways to adjust their existing procedures to remind them to take other users into account when working on projects. They may create checklists or similar tools.

Agencies must also establish a formal procedure for handling any exceptions that may have been included in the policy. This procedure must include high-level sign-off on a compliance document (as stated in the USDOT Design Guidance).

Re-writing design manuals: Note that the USDOT Guidance encourages a re-write of the *primary* design manual, and it suggests that the creation of separate bicycle-pedestrian manuals is only an interim step. A number of jurisdictions have created new design manuals that your agency can use as a model. The Transportation and Air Quality Collaborative in Sacramento, California is notable for developing 'best practices' guides for bicycles, pedestrians, transit—and a separate 'complete streets' best practices guide for putting them all together.

Training: The USDOT Design Guidance recommends "intensive re-tooling and re-training of transportation planners and engineers with the new information required to accommodate bicyclists and pedestrians." Training has already been a valuable outcome of existing policies. For example, California's Deputy Directive 64 inspired a series of trainings for engineers and the Palmetto Cycling Coalition is working with the League of American Bicyclists to plan trainings for South Carolina DOT personnel. You can help your agency connect with a number of organizations and consultants that offer bicycle and pedestrian training courses. Thunderhead organizations can also offer assistance by helping organize trainings (make sure to charge market rate consulting fees) to educate agency employees on implementation issues.

You may also simply need to push the agency to publicize the new policy.

New outcome measures: The best way to test these policies would be to look at what is happening on the ground. However, the most common answers to questions about outcomes in the survey were that it is just too soon to tell if the policies have succeeded, or that no records were being kept. Disappointingly, few localities are collecting any information about outcomes, whether you define those outcomes in terms of roads 'completed,' increases in walking or bicycling, or decreases in crashes. Even in exemplary Oregon, statistics are few at the state level. Bicycle and Pedestrian Program Manager, Michael Ronkin, observed that the state experienced a slight decline in bike/ped commuting from 1990 to 2000, but less than the rest of country; and that crashes are lower than other Western states. He also observed that statistics are extraordinarily difficult to keep. Thunderhead's Benchmarking Project that gathers and compares bicycling and walking data sets from across the country is designed to be a valuable tool in assessing the effectiveness of these policies. This is the only such measurement project that strictly adheres to government endorsed data sets that are uniform across all states. For more information on Thunderhead's Benchmarking Project see: www.thunderheadalliance.org/benchmarking.htm .

An evaluation of the actual effectiveness of the policies included in the survey has not yet occurred. More investigation is needed on the impact of these policies and how to make them work. Thunderhead leaders indicated that even if their policy was not well implemented, it provides additional leverage in advocacy efforts. For example:

> *Internal [CalTrans] allies have seized momentum created by DD-64 to institute a series of bike/ped design trainings for DOT planners and designers.*
> (California)

While few of the current complete streets policies have any sort of metrics, our RECOMMENDATION is that you try to get them included in yours. A very important element of future campaigns will be to include progress indicators or outcome measures, especially those that will easily plug into Thunderhead's Benchmarking Project.

Very few existing policies make any serious attempt to measure new outcomes from the transportation planning process. In fact, most don't even require measuring such conventional outcomes as crash statistics. However, we need these types of measures to document change and to create accountability. Here are a few brief suggestions:

- A new measurement system has been developed in Florida, where planners are using multi-modal level of service (LOS) to measure system quality. Details can be found at www.dot.state.fl.us/planning/systems/sm/los/default.htm.

- A National Highway Cooperative Research Program project on multi-modal LOS is due out March 2005. For details, please see www4.trb.org/trb/crp.nsf/All+Projects/NCHRP+3-70.

- However, don't think that a measurement has to be complex. The Thunderhead Benchmarking Project compares basic statistics about the bicycling, walking, and health environment and will serve as a national measuring tool for all complete streets policies. The League of American Bicyclists' Bicycle-Friendly Communities program also asks for basic statistics.

- Another approach is to create performance goals oriented to the end user, such as, "Can every child safely walk or bicycle from their home to the neighborhood school?"

Staying in close contact while the agency is setting up procedures could make the difference between a good policy, and one that does little to change the status quo. Be sure you have energy, time and resources ready for this stage.

Thunderhead leaders can influence the internal implementation process through a formal advisory committee, or through informal relationships with agency staff. Thunderhead leaders who respect the agency's process can position themselves as a resource, helping bring agency officials' attention to the growing number of documents available to help them implement complete streets.

Making Change on the Ground

Once procedures have been set, the next step is seeing the policy in practice. Continued challenges mentioned by survey respondents included budget issues in regards to projects already underway, right of way acquisition (or lack thereof, also regarding projects underway), public opposition, and tension between different agencies.

As a relatively new concept, we are still learning how to ensure that complete streets policies operate 100% effectively. And unfortunately at this point, little can be learned from the limited number of jurisdictions with policies as few of them are making any meaningful attempts to measure their success. As Complete the Streets campaigns mature, Thunderhead leaders will play a vital role providing important insight on what does work to move complete streets policies from paper to pavement, and what does not.

More implementation ideas can be found in some of the complete streets policies listed in Appendix D.

4

Campaigns
(blueprint for success)

Introduction

While this Guide focuses on complete streets campaigns, this chapter provides a blueprint for crafting and winning any kind of bicycle and/or pedestrian advocacy campaign. In each of seven basic elements of successful campaigns, this chapter will provide some core principles of effective campaigning to help you make the right choices at the right time—the heart and art of strategic campaigning.

All campaigns revolve around the central idea of engaging people to create change. The best campaigns also build your organization on the way toward winning your campaign, so that subsequent campaigns can tackle bigger and bigger issues.

Underlying the core principles for effective campaigns are some basic human truths. One is that while our campaigns are often about very serious matters, they can benefit from a sense of playfulness and humor. After all, what we are working to do is reduce the drudgery of auto dependence and increase bicycling and walking, decidedly more fun modes of transportation! As we work against the formidable opponents of inertia and the status quo, humor can help us have fun and be creative and flexible.

Campaign Examples Description

Here are three campaigns from three different regions of the United States that illustrate the wide range of possible goals, strategies, and outcomes. See Appendix A for details of these campaigns from Marin County, California; Texas; and Columbus, Ohio. Use these three campaigns as encouragement and as a creative starting point in developing the unique campaign that will flourish in your situation. And remember that your campaign will provide inspiration for those who come after you (so please keep good notes!).

Marin County, California: The Marin County Bicycle Coalition worked for six years for a local transportation sales tax election item and finally won its placement and passage on their November 2004 ballot. Measure A, a half cent sales tax increase will generate approximately $331 million over the next 20 years, includes a complete streets policy, contains funding for bicycle and pedestrian enhancements in each of their four key strategies to reduce congestion and improve transportation choices, and provides $36 million for Safe Routes to School for Marin County.

Texas: The Texas Bicycle Coalition (TBC) passed statewide legislation in June 2001, establishing the Safe Routes to School Program administered by the Texas Department of Transportation. Upon the bill's passage, TBC entered phase two of the campaign where even more vigilance was needed to write the rules over the next year and then to re-ignite the grassroots support to successfully demonstrate the need and demand for such a program when the first Call for Projects was announced in August 2002.

Columbus, Ohio: The Central Ohio Bicycle Advocacy Coalition (COBAC) proved successful in their campaign to get the Mid-Ohio Regional Planning Commission (MORPC), a Metropolitan Planning Organization (MPO) for the Columbus region in central Ohio, to adopt a complete streets resolution with detailed policy requiring the routine accommodation of bicycles and pedestrians in the planning and design of all proposed transportation projects using MORPC-attributable federal funds. The entire process took almost 15 months, starting with the first letter to MORPC and ending with a signed Resolution in July 2004.

So, you are all fired up about an issue in your local community, your region, or your state. What happens next?

Thunderhead Alliance Campaign Planning Blueprint

Successful campaigns are well thought out in advance and organized around a clear message and specific goal. Thunderhead's Seven Elements of Successful Campaigns are the basis of our proven Thunderhead Training Curriculum and will help you "keep your eyes on the prize."[1]

1. Issue Focus: Selection and Definition

2. Organizational and Campaign Goals (Short-, Medium-, and Long-term)

3. Resource Assessment

4. Strategic Targets

---

1. Thunderhead's Campaign Planning Blueprint is adapted for bicycle and pedestrian advocacy organizations from similar campaign planning models developed by the Sierra Club, the Midwest Academy, and other environmental and social justice advocacy organizations.

5. Communication

6. Tactics & Timelines

7. Budget and Fundraising Resources

Element One includes how to determine what's an appropriate issue to organize around, and how to define it in crystal clear language so that everyone understands its importance. Element Two emphasizes clear goals, both for winning the campaign and just as importantly, building the organization. The "Resource Assessment" element helps you plan a campaign appropriate to your organization's strengths and weaknesses. Selecting strategic targets is key to staying focused; don't waste your time trying to persuade irrelevant parties. Element Five helps refine your communication strategy. Element Six is the "to do list" of your campaign: which tactics will you employ when. Finally, Element Seven helps you make sure your organization comes out of the campaign stronger financially than it began.

Much like building a house, creating a solid foundation for your campaign will pay off directly and indirectly in the years to come. Planning your campaign carefully reaps the following benefits:

- Your volunteers and allies can be patient, confident that the tasks and incremental victories during the campaign point to the eventual victory they are working for.

- You'll be in the driver's seat, instead of reacting to other's moves.

- You'll be prepared to build your organization through your campaign.

- When you react to opportunities, you'll be more able to stay on message and more prepared to adjust your plan to affect victory as well as build your organization.

Getting Started with a Campaign Planning Meeting

Before jumping in and starting, we strongly recommend you launch your campaign effort with a planning meeting. Invite key stakeholders, especially people who will lead various aspects of the campaign. At your meeting, you'll develop your campaign plan element by element. Use the Campaign Planning Workbook (available in the members section of thunderheadalliance.org or by emailing us for a copy) to guide your planning. The next section of this Guide details the process for completing each element.

When you're finished with the Workbook, someone needs to distill the information onto the Campaign Blueprint (Appendix B and also available in the members section of thunderheadalliance.org) for sharing with stakeholders. This written plan is your shared understanding of the campaign goals, messages, and strategy. It is easy to get side tracked during a long and emotional campaign. The plan serves as a reminder to the whole team of your direction and focus for the length of the campaign—it is your blueprint for success!

Here's a checklist to help you make sure your campaign planning meeting is a success:

<u>Setting:</u> Find a quiet, comfortable space dedicated to this meeting without distractions such as small children, cell phones or non-participants.

<u>Supplies:</u> Have good-sized chalk or marker board or a way to mount large sheets of paper for recording proposals, and include large bright markers, and note pads and pencils for personal use.

<u>Attendees:</u> Invite your Board members and staff as well as key stakeholders in the potential campaign.

<u>Time Frame:</u> Set aside 3-5 hours, either at once like a "retreat" or over two meetings. Always begin and end on time! This encourages focus and commitment.

<u>Preparation:</u> Have handouts ready with background information on the campaign issues as well as the 7-step planning blueprint that you will be working with in the next part of this Guide. Copies of the Workbook would be very useful. Also provide everyone a list with each participant's phone and mail contact info.

<u>Roles:</u> Use a designated **facilitator**, perhaps someone from outside the organization with experience managing organizations. Their job is to tactfully ensure that everyone has a chance to comment and to enforce the ground rules. Also designate a **recorder** who takes thorough notes of the discussion and perhaps a **timekeeper** to keep the meeting flowing smoothly so that all the topics can be addressed.

<u>Rules:</u> Open the meeting with the facilitator establishing the ground rules. These might include the order in which discussion will occur, time limits on speaking, no interrupting, no repeating or rephrasing, and staying on topic.

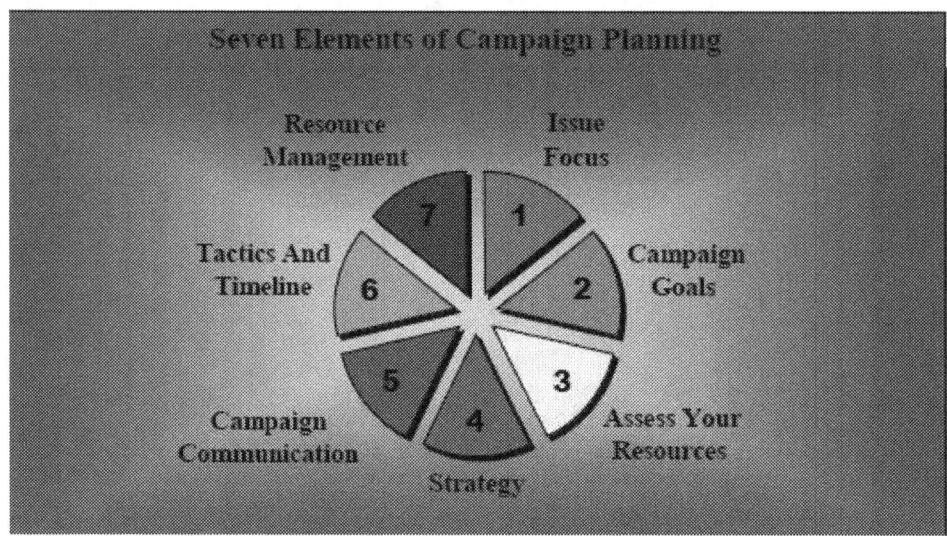

Seven Elements of Successful Campaigns

In this section, we discuss each of the seven steps of the campaign planning process and show how the three campaign examples incorporated each of the elements into their respective local and state-wide campaigns.

Element 1—Issue Focus: Selection & Definition: The first step is selecting your campaign issue. Then, you must be able to define it clearly so that it can be expressed and understood in a sentence or two.

Selecting Your Issue. If your issue does not score well on Thunderhead's Campaign Checklist below, it's probably not a good issue for your organization. Seeing how well complete streets policy campaigns score on the checklist, you will understand why the Thunderhead Alliance prioritizes getting complete streets policies passed throughout North America.

1. Does it align with model campaigns?	Yes, thanks to this Guide.
2. Is it winnable?	Yes. Some communities already have complete streets policies.
3. Will the campaign result in a definite and quantifiable improvement in the community (i.e., will it increase bicycling and/or walking and reduce crashes)?	Yes, complete street policies result in permanent change, and benefit both walking and bicycling.
4. Does it set long-term improvements to the bicycle and pedestrian environment?	Yes, extremely! Physical improvements to the streets are definitely long-term, and complete streets policies accomplish long-term changes in planning as well.
5. Enlists the involvement of important groups of people.	Yes. For example, groups like AARP, school safety groups, and transit advocates may all be important parts of a complete streets coalition.

6. Does it fit your organization's mission and culture? Does it unify and not divide your constituency?	Yes. All bicyclists, from recreational tourists to daily commuters, and all pedestrians would benefit from a complete streets policy. Providing safe streets for walking and bicycling is not controversial in any community.
7. Involves your current members in a meaningful way.	Yes. Lobbying, bringing in coalition partners, researching data about the current state of incomplete streets are all ways in which current members can help the campaign meaningfully.
8. Will it attract new members.	Yes. This is a sufficiently sweeping change to entice many people to join the organization leading the effort.
9. The issue is both broad and deep: many people care about it and some are very passionate.	Absolutely. You will find ardent supporters among those who support smart and efficient government investments in sustainable transportation, and supporters across the entire social spectrum.
10. Builds organization's political power.	Yes to a degree, depending on the type of complete streets policy. Policies achieved through quiet agreements at the staff level build less political power than policies achieved through legislative votes. Still, either type of policy builds important political power.
11. Will leverage positive media and promotion of your organization.	Absolutely. Nobody wants an "incomplete" street, and everybody will appreciate a positive organization seeking to make sure our communities are safe and complete!
12. Has strong income potential.	Yes to a degree. Members and donors are most likely to support complete streets policy campaigns if they understand how it will help them in their everyday life.

As an exercise in defining your issue, the first step is to articulate the issue in clear concise language that a child can understand. Every one of these four parts is essential:

- Identify the problem.

- Formulate the solution.

- Illustrate how to implement the solution.

- Show the various roles people can have in the solution.

In Marin County, California the issue was: "Traffic is bad and getting worse. The public wants more places to walk and bike safely." They were also ready to hear about other solutions, especially since the efforts to pass the half-cent transportation sales tax had failed on three other attempts (in 1980, 1990 and 1998). The general public was ready to participate and give their input into how to fund transportation alternatives while considering the needs of bicyclists and pedestrians.

In the Texas Safe Routes to School campaign the issue was: "Schoolchildren are not getting enough healthy exercise and traffic congestion around schools is burgeoning; safe biking and pedestrian facilities in and around schools which encourage self-reliant transportation will relieve both situations." The political climate was right and TBC had good model legislation borrowed from the successful passage of the California Bicycle Coalition and Surface Transportation Policy Project's Safe Routes to School program in 1999.

In Columbus, Ohio the issue was: "A review of the MORPC Transportation Improvement Plan (TIP) reveals almost none of the projects approved includes accommodations for bicycling and walking, as required by federal transportation law and policy." COBAC knew they could not wait another three years before MORPC unveiled a new TIP to begin to change bicycling policy.

Element 2—Campaign Goals: Specific tangible goals must be defined. What will victory look like; what will signal the end of the campaign? Some campaigns may last for years so it is important to have benchmark successes: short-, mid-, and long-term goals. Incremental progress toward the ultimate goal helps keep campaign workers upbeat and enthusiastic. Stay flexible and roll with the situation, celebrating smaller victories while keeping the long-term goal in mind. Always have SMART goals and objectives: Specific, Measurable, Achievable, Realistic, Timebound.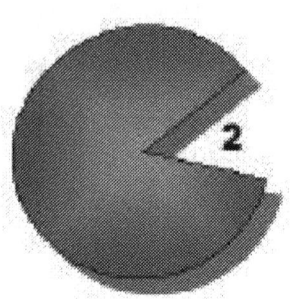

Campaign goals can be subdivided into two categories: issue goals and organizational goals.

Marin County, California Issue Goals: work bicycling and walking programs into each of the four strategies in the expenditure plan, and for all funded projects to consider the needs of bicyclists and pedestrians. The overall goal for the transportation sales tax was to develop a plan to increase mobility and reduce traffic congestion.

Texas Issue Goals: Phase I: passage of House Bill 2204 directing the Texas Department of Transportation to establish the Safe Routes to School Program. Phase II: create a groundswell of public demand on a statewide level, prior to the first Call for Projects announcement and during the four-month application period.

Columbus, Ohio Issue Goals: seek a Resolution from the Metropolitan Planning Organization to require routine accommodations of bicyclists and pedestrians in the planning and design of all proposed transportation projects using MORPC-attributable federal funds.

Organizational goals will need to be developed to strengthen your organization over the course of a campaign. To be deemed a success, every campaign should leave the organization larger and stronger than when it began. As with issue goals, these should be written down, made accessible to all campaign leaders and monitored throughout the campaign to be adjusted as necessary. Sample organizational goals might look like the following.

- Nurture two new leaders within the organization willing to take on specific responsibilities.
- Establish good working relationship with two new coalition partners.

- Create a database of contacts at 2 local TV stations, 3 radio stations, 3 newspapers, and 4 magazines.

- Net $10,000.

- Spend at least half of campaign's time and effort reaching out to non-members.

As with issues, stay specific so that you know when you have achieved the goal, and be sure to acknowledge and celebrate every victory.

Marin County, California Organizational Goals: MCBC set several organizational goals. 1) Participate in all large-scale transportation planning in the county to ensure bicycling and walking are included and to incorporate this concept into all fundraising appeals and year-end donations. 2) Nurture the leaders on the Citizens' Advisory committees to create new spokespeople for the cause. 3) Teach existing MCBC members how to talk to their representatives and the decision makers from the County Board of Supervisors and each of the 11 cities and towns. 4) Align the MCBC with the business and environmental community to make clear MCBC's invaluable role in the passage of the transportation sales tax. Every public works director and public official knows that "the bicycle lobby" led by the MCBC created the momentum for the campaign.

Texas Organizational Goals: TBC used the Safe Routes to School campaign as a way to build their relationship with the Texas Department of Transportation, schools, city planning departments, neighborhood associations, bike shops, and bicycle clubs. Because of the state and federal legislation and agency support, the program proved to be a good common ground and source of infrastructure funding as well as for education and encouragement funding. The Safe Routes to School campaign was a policy effort that attracted bipartisan appeal.

Columbus, Ohio Organizational Goals: At the start of COBAC's campaign to get MORPC to adopt a routine accommodation policy in April 2003, COBAC was just beginning to revive itself from ten years of inactivity. Rather than waiting to develop the organization before beginning the campaign, the leaders of the COBAC revival decided to proceed with the Complete Streets campaign with minimal resources, . intending to show through a winning campaign the need for and value of bicycle advocacy. COBAC's organizational goal was to emphasize public outreach and to develop long lists of interested potential members to recruit from later.

Element 3—Assess Your Resources: Starting at the initial planning meeting, the group should discuss two key components involved in the campaign: 1. organizational strengths and weaknesses and 2. allies and opponents.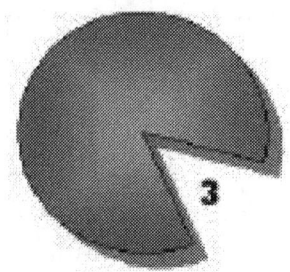

Clearly assess your organization's strengths and weaknesses. What resources do you have? Think in terms of people, money, time, and connections with policy makers. Who in the group has what skills? Who can keep databases up to date keeping track of volunteers, decision makers, and media contacts? Who is an effective public speaker? Are there any lawyers, researchers, press release writers, or fundraisers? Does anyone have influence with potential allies or donors? What resources do you lack? Do you have sufficient information about the processes you hope to affect? Do you have compelling data that backs up your case for wanting this issue in the first place?

After measuring your organizational pluses and minuses, you will have a better idea of what allies to seek. Besides sheer numbers of support, you will want to persuade people who can compensate for your weaknesses. It is important to cultivate and include as many allies as possible.

Your best chance of success will be to work with a broad-based coalition of special interest groups or community organizations whose interests have some overlap with yours.

Getting to know the opposition is equally important. Who isn't interested in your project or is actively opposed to it? What are their reasons? Perhaps in listening to their concerns, you can alleviate them and find common ground. Perhaps they have a misconception that you can clear up. It is vital to treat opponents with respect and attempt to work with them. Think of every opponent who could possibly benefit from your issue—and then convince them of it.

Marin County, California Resources: Once the transportation sales tax expenditure plan was created, more than 100 organizations joined as allies to pass Measure A with 71% of the vote. Supporters included the Association of Realtors, the Commission on Aging to the Sierra Club, Greenbelt Alliance, the Builders Association, and Transportation Alternatives for Marin, making it an unprecedented example of environmentalists and businesses working together. The only opposition was from the Taxpayer Union. Another one of MCBC's organizational strengths was knowing that the Marin County Supervisor who Chaired the Transportation Authority of Marin was a leading champion and visionary on how to integrate the various modes of transportation.

Texas Resources: TBC's strongest allies were the statewide associations and 80 bicycle retailers who endorsed the bill's passage and subsequent grassroots campaign to create overwhelming demand for the Safe Routes to School project applications. The groups included the Texas Medical Association, Texas Congress of Parents and Teachers, Texas Hospital Association, and the Texas Association of Health, Physical Education, Recreation and Dance. During the four-month initial project application period, the Department of Transportation received more than 300 applications for a total of $45 million in project requests for the $3 million program.

Columbus, Ohio Resources: The supportive, sympathetic staff at MORPC proved to be the biggest ally. There was no organized opposition except from two engineers at two different counties who objected to language that could have been interpreted as setting a minimum percentage to be spent on accommodating bicycling and pedestrians in each project. The COBAC president served as chairman of the bicycle citizen advisory committee, which helped him stay connected with the MPO plans, process and staff.

Element 4—Strategy: Once you've defined your goals, selecting the strategic vehicle for achieving those goals is next. Part of choosing a strategy is to define who you will target with your campaign: lawmakers, policy makers, and/or voters and what methods you will use to influence them.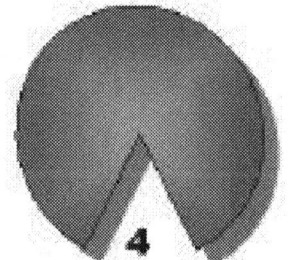

For example, if your issue focus or campaign is "Complete the Streets" and one of your goals is to adopt a policy accommodating bicyclists and pedestrians in all transportation projects, you might try to address this challenge in several ways. You could get a resolution passed at the local or county level. You could work with your city, MPO, or state DOT to adopt an internal policy. You could launch an intensive state level legislative campaign. Or you could simply set the stage for a future campaign through a PR campaign.

What is most important is that you select the most appropriate strategic vehicle given the current political climate, relationships with other groups, and an honest assessment of your organization's resources. Examples of other strategy elements might include, lobbying lawmakers or pressuring public agencies.

After you define your strategic vehicle, you will need to figure out exactly what individual(s) you need to target in order to claim victory. Carefully selecting your targets helps you focus your efforts on the right individuals. Avoid those who you will not be able to persuade or who are not so important to the final decision maker and don't waste energy on those who already support you. Your targets can be divided into *primary, secondary* and *public* targets.

Your PRIMARY targets are the decision makers. Who has the power to make decisions and deliver a victory for you? This question should be answered with a specific name of an individual and not merely the name of an institution or governing body. What if you are not sure whom to target? Find out everything you can about how the governing body works. Follow-up by researching all you can find about its various members such as their past voting records, political connections and past positions on your issue. In some instances, if you need to get a majority vote on a particular issue from a group of individuals such as the legislature, board of directors or planning commission, then you will need to select a subgroup among this governing body to target.

Your SECONDARY targets are those people who can influence your primary decision makers. Would they be willing to use their connectedness to your primary targets to advance your goals? Think about what you can ask these various individuals to contribute to your campaign.

The last stage of developing your strategy is targeting the PUBLIC audiences. Take a look at the community or state in which you are waging this campaign and determine which specific groups of people you can enlist to create demand and hold the decision makers accountable for meeting the demand. It is important to select no more than two or three public audiences. You can then focus all of your attention and efforts on persuading these people to join your campaign and not waste any resources on individuals who are not in the targeted group.

Marin County, California Strategy: MCBC had a two-part strategy. The first was to get bicycle and pedestrian elements included in each of the four strategies that were designated in the transportation sales tax expenditure plan. Their primary targets were the members of the Transportation Authority of Marin, the commission who held final authority over the structure of the sales tax expenditure plan. decision making body for the elements in the plan included representatives from the County Board of Supervisors and each of the 11 cities and towns. The second strategy was helping get the transportation sales tax plan approved by two-thirds of the voters.

Texas Strategy: TBC also had a twofold campaign by initially targeting state lawmakers to get the bill passed, and then in the ramped-up grassroots second phase, addressed statewide PTA's, school superintendents, teachers, mayors, and city managers as secondary targets with the Department of Transportation as the primary target.

Columbus, Ohio Strategy: COBAC used the leverage of federal regulations to make MORPC the primary target and demand that they withhold all federal funds from all projects listed in the TIP that did not include bicycling and walking.

Element 5—Campaign Communication: Now it is time to determine what you are going to say. The most important thing to remember is "respect the viewpoints of the people you wish to persuade." Find out what is important to them and couch your message in terms they can understand and that have relevance in their lives. Be certain that your position is factual and well documented. Ideally, your communication strategy is a diverse one, including a slogan, a 30-second "stair speech," a personal story, and a 150-word letter to the editor.

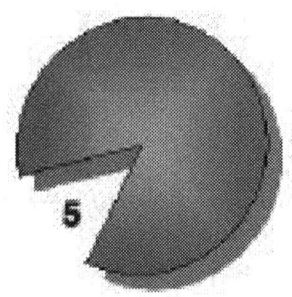

Your slogan should be clear and concise: 10 words or less. "Complete the Streets" is an example.

Your "stair speech" is a concise and compelling story that can be expressed in 30 seconds, the time it takes to ascend one level of stairs with a policy maker or newspaper editor. A stair speech has a "hook," something to grab the listener's attention; a quick statement of the problem you're trying to fix; the solution; and what the listener can do to implement the solution.

Your story can put a human face on your policy campaign by relating it to a real person locally that people can identify with. It can be expressed in public testimony at meetings, in your own newsletter, and at public gatherings. Your story can be emotional, and can take a few minutes to tell.

Finally, you should be able to express your issue clearly in a letter to the editor of 150 words or less. The "letters to the editor" section is among the most-read sections of the newspaper.

Working on the above elements of your communication strategy as part of your campaign plan will help all campaign workers stay on message during the campaign.

The media can be a highly effective ally in your communications strategy. Target local newspapers with human-interest stories, letters-to-the-editor, and op-ed pieces. Cultivate relationships with sympathetic reporters or editors. Provide factual, concise, and interesting material. And persist. If a story is not run, politely check with the paper or magazine and find out why. Ask what you can do differently to ensure coverage of your issue or event.

Marin County, California Communication: MCBC's message was "bikes are part of the solution!" They organized letters to the editor efforts among supporters and kept their members informed through newsletters and their website. They trained parents to speak on congestion management, recommending bicycling and walking as alternatives to being driven in motor vehicles. MCBC used the voter polls to their advantage by showing the existing public support for Safe Routes to School and bicycle and pedestrian improvements. They provided easy-to-use sound bite quotes for the media on a regular basis. They set up meetings with members of the Transportation Authority of Marin to make their case (bringing local constituents to the meetings). They stayed consistently on their message.

Texas Communication: TBC provided testimony before legislative bodies in passage of the bill while supporters generated hundreds of original letters to demonstrate support and demand during the Rulemaking phase. TBC sent 10,000 letters announcing the Call for Projects applications to school superintendents, city managers, and PTA members. More than 250 bike shops distributed posters and postcards announcing the application process. Prior to the official Call for Projects announcements, TBC visited all 180 members of the Texas legislature with a pre-written announcement and sent press releases to more than 450 Texas newspapers generating more than 300 articles about the Call for Projects.

Columbus, Ohio Communication: The use of rhetorical arguments, public health information, and sample text from other jurisdictions in all correspondence and testimony helped COBAC stay consistent and on message.

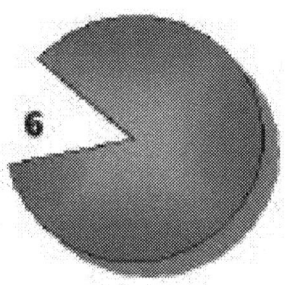

<u>Element 6—Tactics and Timelines:</u> By this time, you have developed your issue, goals, resources, strategy, and communication plan. The next step is to develop the campaign's tactics and timelines, the means by which you will carry out your strategy. Too many organizations make the mistake of starting their planning by brainstorming a list of activities—jumping right to a "to do" list—rather than doing the research and analyzing their issue, resources and targets for accomplishing their goals. Only once you've finished your campaign plan and considered all your tactics, should you launch your campaign with the first tactic: a bold public announcement kicking off your effort.

Staying true to the messages developed in the previous section, develop tactics to reach your strategic targets with a demand for your goal. There are three general types of tactics you can employ to create demand: direct contact, general visibility, and media tactics.

Direct contact tactics are aimed at your selected public audiences and focus on personally engaging people in the campaign. Direct activities might include high-level meetings, phone banks, door-to-door canvassing, neighborhood coffee or house parties, leafleting, petition signatures, and/or written communication.

Another way to create more demand is through general visibility tactics, directed toward all of your targets, generating a community buzz. Whereas the direct contact activities listed above are like spraying a garden hose full force at a particular target, visibility activities are like a lawn sprinkler, covering more ground but with much less intensity. Visibility activities might include rallies, demonstrations, yard signs and other materials such as campaign buttons, t-shirts, and bumper stickers that communicate the campaign message.

You will want to inform, involve, and connect people with ways to be part of your campaign. Once you have a support base, use it to generate media tactics coverage with rallies, hearings, and events. Use direct contact methods to build a support base and then to contact legislators, agency members, and everyone that you have identified as a target.

Make your campaign irresistible with a growing swell of interest. Perseverance and persuasion are your best tools no matter what tactics you choose.

Marin County, California Tactics: MCBC staff attended every meeting of the Transportation Authority of Marin for six years to serve as the voice for bicyclists and pedestrians and to get bike/ped elements included in the expenditure plan. They created a Position Paper which outlined their goals and the exact amount of funding they desired in each of the four strategies presented in the plan. MCBC showed how bicycling and walking improvements would improve mobility in their community and provided sample language to be included in the expenditure plan. The Paper was heavily circulated to educate members of the Transportation Authority and the five Citizens Advisory Committees which were charged with making recommendations on funding levels. MCBC waited until the expenditure plan was approved (with good provisions for walking and bicycling) before signing on as an endorser of Measure A.

As a Phase II strategy, a key tactic was to show the power of the bike lobby for being a team player in getting the sales tax approved. Thus, immediately preceding the vote, MCBC organized rallies of school children and people with disabilities to garner positive media attention and to convince voters to pass the measure. They also developed e-mail alerts, created downloadable posters on their website, and held phone banking out of their office. Election day morning found forty volunteers (organized by MCBC) positioned at strategic freeway entrances holding signs saying "Yes to Measure A." Note how this tactic not only supported the *issue goal* of passing the measure, but also the *organizational goal* of positioning the "bicycle lobby" led by the MCBC as an important player in county politics.

Texas Tactics: TBC prepared well before the start of the legislative session by hiring a professional lobbyist, along with a volunteer campaign consultant and a volunteer campaign manager. Several of the largest bicycle retailers allowed TBC to use their customer list and cross check it with an enhanced voter registration list to identify bicyclists and send targeted Call-to-Action requests. Hundreds of targeted letters were sent to the transportation commissioners demonstrating the popularity of the Safe Routes to School program. Targeting specific TBC members through email and their website prevented list fatigue. A new impression of bicyclists emerged by asking attorneys, doctors, engineers, and even a 12-year-old girl who started a Safe Routes to School petition at her school to testify at committee hearings.

Columbus, Ohio Tactics: COBAC attended official meetings, submitted comments, encouraged testimony at public hearings, worked the political process, and helped write and revise language. COBAC submitted a letter to MORPC in April 2003 demanding that all federal funds be withheld from all projects listed in the TIP that did not include bicycling and walking as required by federal transportation law and policy. In May 2003, COBAC sent a letter to the Federal Highway Administration with objections regarding Ohio DOT's 2004-2007 STIP, copying Ohio DOT.

Once you have finalized all of your campaign tactics, construct a timeline. Using a regular calendar, write in all of the campaign activities and draw lines to indicate exactly when they will take place. Next to the activity, indicate the specific individual who will be responsible for each activity. This last step will help you to evaluate if you have enough people to cover all of your tactics. No doubt, you will need to produce updated "to do" lists on a regular basis to keep track of things. This initial timeline will prove valuable and give the entire team a shared understanding of the overall pace of the campaign and what lies ahead.

Marin County, California Timeline: MCBC embarked on their campaign the day after the 1998 transportation sales tax failed. At that time, they had no idea that the process would last for six years.

The real campaign to pass Measure A kicked off in August of 2004, after the expenditure plan had been finalized, which was three months prior to the election.

Texas Timeline: TBC spent more than two years in developing the legislation and building coalition partners prior to introducing the bill in January 2001 for the 77th Texas Legislature. The session runs 140 days, and the bill was not passed until the last day of the session in May 2001. The Rules Adoption Process with the Department of Transportation took almost one year, from July 2001 to July 2002. The announcement for the first Call for Projects applications was made in August 2002. TBC was highly engaged during the application period since the deadline to submit project applications was December 2004. The different phases of the campaign totaled almost four years with two years for pre-planning, six months to pass the legislation, one year to write and adopt the rules, and four months to promote and encourage communities to submit applications.

Columbus, Ohio Timeline: Between April 2003 when objections to the MORPC TIP were lodged and when MORPC adopted the routine accommodation policy on July 22, 2004, approximately 200 hours over 15 months were spent on advocating for this policy.

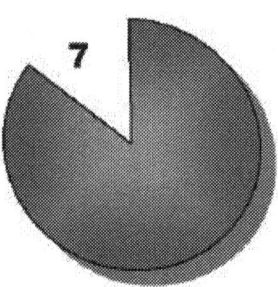

Element 7—Resource Management: Campaigns cost money and they need people. You will need to develop a campaign budget, find donors, undertake fundraising, and seek and manage volunteers (and staff).

Develop a budget including a fundraising plan right from the beginning. It is one of the single most important papers in a campaign. This document will set forth exactly how you will invest your energy and how you will not. Know the costs of proposed materials, phone calls, postcards, yard signs, professional lobbyists, lawyers to review documents, and don't forget your overhead costs including staff time. Assume that the campaign will run into unexpected expenses and build at least a 10% contingency fund into your plans.

In fundraising, look for donations of cash as well as in-kind services. Create a written monetary plan that states your overall financial goal as well as goals by source: organizational support, individual donations, events, and perhaps grants. To create your budget use an assessment of your organizational resources, your ally's resources, and the cost of implementing your tactics.

After you finalize your fundraising plan and budget, create a cash flow chart that shows how much money will be coming in and when, and how much money will be going out and when. The cash flow should be monitored and adjusted on a regular basis.

Managing people, recruiting volunteers, and maintaining a volunteer base are as critical as the fundraising and budget work. The most effective tool in recruiting and maintaining volunteers is easy: Informing, Involving, Thanking, and Asking. Brainstorm on how to attract volunteers. Have jobs available when people ask about helping. Start them with small discrete tasks allowing them to take on more as their confidence and enthusiasm builds. Have written sets of instructions on how to complete tasks. Make volunteers feel at home and appreciated within the organization. Learn first names, ask for suggestions and thank your volunteers again and again and again. Take the time to nurture leadership qualities. Have plenty of food and drinks on hand. Some organizations even go out of their way to keep the favorite foods of their "Super Volunteers" always in stock. Volunteers will stay longer and most likely come around more often if they can count on food and beverage as part of the volunteer perks.

Campaigns regularly suffer loss of personnel through burnout or the normal changes that occur in people's lives. Grow new leaders for your organization by keeping an eye out for leadership qualities. Many people need encouragement to take leadership positions—your job is to give it to them.

Whether you're talking about money, volunteers, or political connections, strengthening these resources happens in the same way: Informing, Involving, Asking, and

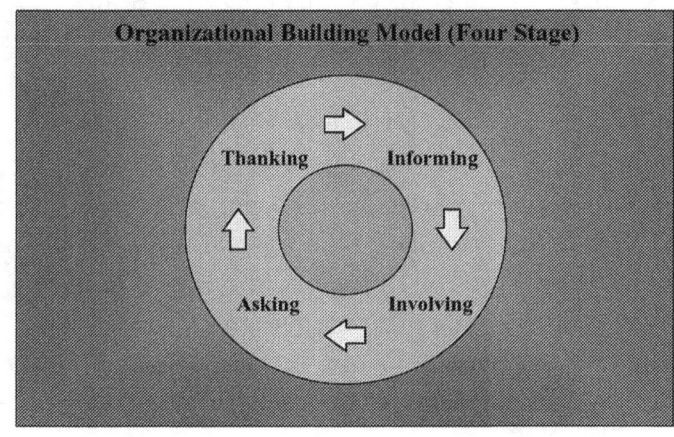

Thanking (again and again). And don't forget to add elements of fun wherever possible to hold volunteers together during the inevitably discouraging process of making effective change in the status quo.

Marin County, California Resource Management: MCBC funded the Measure A campaign out of their normal budget. MCBC coordinated more than 100 volunteers during the six-year campaign and over 40 volunteers were recruited the day of the election. In their membership solicitation and fundraising materials, MCBC did indicate that the organization was working to ensure that the needs of bicyclists and pedestrians would be included in whatever transportation sales tax was sent to voters.

Texas Resource Management: TBC raised more than $80,000 to fund the legislative session and subsequent encouragement campaign to create high demand for the first Safe Routes to School application period. The fundraising was done among bike shops, individual donor direct mail appeals, bike shop customer lists, on-line donation capabilities, and a $10,000 grant from Bikes Belong. More than 100 cyclists attended the 1st Cyclists in Suits Bike Lobby Day at the Texas Capitol and listserve managers from 20 different bike clubs and websites forwarded action alerts to their members.

Columbus, Ohio Budget and People Power: COBAC spent approximately 200 volunteer hours getting the MORPC Resolution adopted, leading to a natural segue for the next campaign by asking the same questions of Ohio Department of Transportation and why they are not funding more bicycle and pedestrian projects as part of their State Transportation Improvement Plan.

Blueprint for Success

You have now gone through the "7 Elements" and can imagine the point at which you will have essentially organized your campaign blueprint—your written plan which describes the Who, What, When, Where, Why, and How of your entire campaign.

See Appendix B or email us for your own copy of Thunderhead's Campaign Planning Blueprint. Fill it out for your whole team to reference.

courtesy of David Crites

Now it is time to go out and do it. If you follow your plan, staying focused on your strategy and your message, you have every chance of success. Be persistent, work hard, and have fun! As you move forward keep Thunderhead contact info close. Though your plan may seem clear now, you will likely come across unexpected turns. Call or email us anytime. We are here to help.

You may ask, "How are each of the three campaign examples doing now?" Here are current snapshots of each campaign in sum. You can get a flavor of their unique blueprints and how they are faring now. And don't forget that much of this information is offered in yet another format in Appendix A.

Marin County, California: The Marin County Bicycle Coalition committed to working on the passage of the local transportation sales tax the day after its defeat in 1998. Polls done on why the previous versions had failed indicated that voters wanted more specifics on how the money would be spent and exactly what improvements would be implemented. MCBC began its campaign by showing officials that the County needed an updated Bicycle and Pedestrian Master Plan. The previous version was done in 1974 and it was not being implemented. Public outreach for developing the Bicycle and Pedestrian Master Plan took place in 1999 and 2000, with the final plan being adopted in 2001. Simultaneously, the County conducted detailed studies on the needs and future operation plans for local bus transit and a start-up rail service, high occupancy vehicle lanes, and local roads. The compilation of all of these studies resulted in the Transportation Authority of Marin's creation and adoption of a Transportation Vision Plan in the year 2003. The plan depicted a multi-modal vision for travel, and indicated that the County would need $1.3 billion to fully implement the recommendations for all of the modes of transport. Calculations showed that the amount of funding that would be generated through a 20-year, half-cent, transportation sales tax would be only $331 million. Thus, the Transportation Authority embarked on an extensive public process to determine what projects to include in the future sales tax. They were looking for the highest priority transportation projects that would resonate with voters and increase mobility.

Marin County Bicycle Coalition attended every meeting of the Transportation Authority to make sure that they had a seat at the table. Instead of lobbying for a specific percentage just for bicycling and walking, they lobbied for the needs of non-motorized travelers to be included in each of the four funding strategies in the draft expenditure plan: improving transit, completing the HOV lane, local roads, and school access. The original plan for "local

courtesy of Michael Ronkin

roads" did not incorporate bicycling and walking as routine element of completing the roadway. Seizing the opportunity, MCBC created a Position Paper to explain how the needs of bicyclists and pedestrians could be integrated into each of the four categories and that it would be much easier to incorporate bicycle facilities and widen the roads during the design phase instead of waiting for the next bike-car crash. As a result, the local roads component was expanded to be called "local roads, bikeways, and sidewalks" and it was stipulated in the legislation that all roadwork must consider the needs of bicyclists and pedestrians. One of the biggest direct benefits MCBC has received is the bragging rights to members by telling them how *their* organization shaped Measure A, the most powerful transportation tax funding mechanism that will come along in 20 years. By showing up at every meeting, they were able to better assess and know when to call out the troops. MCBC stayed consistent with their message by saying "bicycle" at every meeting, while leaving decision makers with a favorable impression that MCBC was a strong team player.

Executive Director Deb Hubsmith encourages other bicycle and pedestrian organizations to get involved if their community embarks on a transportation sales tax process. Her advice is that the biggest payoff comes in the end and is only possible if you stick with it and put forth your vision. Marin County's payoff will come in the form of $331 million for transportation improvements, a complete streets policy, and $36 million dedicated for Safe Routes to School over the next 20 years. Now, a third phase of the plan, is monitoring implementation of the transportation sales tax!

Texas: The Texas Bicycle Coalition's realization came almost instantly after winning the hard-fought battle to pass the Safe Routes to School bill during the 2001 Texas Legislature. They told themselves, "Get ready for the next phase because the real campaign was only just beginning. Forget about savoring the sweet smell of victory. This is no time to rest. Getting the bill passed was just the start." Now TBC had to create an even stronger demand and grassroots support than they did during the legislative session because they were in the Rules Adoption process with the Department of Transportation. The three years following the passage of the bill produced some interesting developments. TBC spent almost as much time, effort, and organizing after the legislation was passed as it did to get it passed. TBC left the legislative phase of the campaign a much stronger organization than when they went in. Long-time Capitol staffers remarked they had not seen a grassroots campaign like the one TBC did in 2001 in over ten years. This kind of recognition increased TBC's credibility and enlarged the pool of organizational and political allies.

TBC capitalized on the threat of an anti-bike bill during the same time period which would have banned bikes from 30,000 miles of roadways as a way to generate continued momentum to pass the positive legislation. At one point during the rulemaking phase, almost nine months after the Safe Routes to School bill's passage without any funding source, the program got a huge boost when the bill's sponsor announced $3 million for the Safe Routes to School program funding. Prior to the official Call for Projects announcement in August 2002, TBC visited all 180 members of the Texas Legislature with a press release and to thank them for voting to pass this momentous legislation for Texas children and neighborhoods. More than 300 daily and weekly newspapers ran an article about the Call for Projects.

Even though applicants only had four months to submit an application, by the end of the December 2002 deadline, the Texas Department of Transportation had received more than 300 applications for a total of $45 million in requests for a $3 million program. In February 2003, the Texas DOT commissioners announced the 27 projects selected and even increased the funding to $5 million because of such high demand.

TBC's positioning as a leader within Texas on Safe Routes to School has resulted in a three-year $1.5 million "Safe Routes to School Program" education and encouragement grant, funded in part by the U.S. Department of Education, through the Carol M. White Physical Education Program. The grant, awarded in October 2004, will help launch a massive bicycle safety, physical activity, and health awareness program, aimed at encouraging children to bike and walk to school in 27 cities and towns in Northwest Texas. The training of 750 teachers, potentially reaching 38,000 fourth-and fifth-grade students in 298 schools, will require TBC to hire thirteen full-time staff in 2005 to serve as local outreach coordinators.

Columbus, Ohio: COBAC spent over 15 months in their quest for the adoption of a complete streets policy by the Mid-Ohio Regional Planning Commission, starting with the first letter in April 2003 and ending with the July 2004 Resolution. COBAC president John Gideon cites several reasons as key to the success of the campaign. Gideon believes he was fortunate enough to have a forward-looking Metropolitan Planning Organization that really did want this policy.

Of course, having a little rivalry among the neighboring states who had already implemented the policy was a plus. Another factor that turned into an advantage was Gideon's position as a member of the MORPC Citizens Advisory Committee. This enabled him to stay connected with the MPO process, plans, and staff. The campaign also provided some direct and indirect benefits to COBAC.

The campaign success demonstrated a direct benefit to COBAC by establishing instant organizational credibility and it cemented the relationship with the MPO and jurisdictions. COBAC is no longer perceived as a narrow interest group. The success has also created a demand and need for this type of advocacy organization in Ohio. The natural transition from having a regional success gave COBAC more leverage to pursue changes at the state level with the Ohio Department of Transportation's policy on accommodating bicycles and pedestrians.

COBAC would also have done some things differently. If they had known MORPC better, they would have phrased the language in the first letter a little less aggressively. COBAC realized the need to include broader coalitions to show strength and would have liked more pedestrian and people with disabilities organizations to participate, but they were not interested at the campaign formation time.

The key to creating and implementing successful campaign efforts is to honor the process. Make sure you account for all seven components of the campaign planning when contemplating which activities to support your goals. If you go through this rigorous process, your campaigns will become stronger and your organization will emerge even each time. Our bicycle and pedestrian issues not only deserve this kind of attention and thoroughness, they demand it. And the Thunderhead Alliance is here to help your campaign efforts.

5

Communications
(a toolkit)

Introduction

Complete streets is more than just a new name for what was once referred to as routine accommodation. The phrase is useful not just as a description of a policy, but also as an independent communications tool. This phrase is active, flexible, and imbeds a fundamental message we want to send: that streets are not complete until they are safe and convenient for travel by foot or bicycle, as well as for transit users, people with disabilities, and people in automobiles. A street without such safe passage is by default 'incomplete.' This puts us a step ahead of opponents who would like to characterize complete streets policies as mandates that are an "expensive special" accommodation. Since most Americans walk, and many bicycle, use transit, or have disabilities, this is an important reframing of the way we view the road network.

Even if you are not actively pursuing a specific complete streets policy, using the term can advance bicycle and pedestrian advocacy. This chapter is designed to help you do that.

The Cost Misconception: A common misconception is that complete streets cost more to build than incomplete streets. In fact, complete streets most often cost no more and many times can cost less than incomplete streets. For instance, a common street cross section that serves only cars is a four lane speedway with no shoulders, sidewalks or intersection treatments for people. Using the same right-of-way width, this design can be reshaped into two narrower through lanes, one center turn lane, and bike lanes and sidewalks on both sides. By using less width for the most expense elements, truck weight standard asphalt and subsurface, and adding less expensive sidewalks, this design, often referred to as a "road diet" when applied to existing roads, actually saves money. Not only that, this design has been proven to improve traffic flow and safety for motor vehicles by better controlling turning movements. Many other complete streets designs offer similar cost savings. You may even want to bring up the economic benefits of streets that attract visitors and offer access to more employees. Be sure to address this misconception early in your campaign so that you can focus your valuable time on instituting a policy for your communities.

When you are discussing bicycle and pedestrian friendly changes with decision makers, talk about remolding the same street materials into complete streets. Consider writing an article for your newsletter explaining the idea to your members, or updating your website. Use the term when speaking with reporters, in written testimony, and in meetings and conversations. In short, you will play a vital role in helping us propagate this term by using it whenever you can. We need this phrase to become the shorthand for our nation's transportation network that truly welcomes people on foot and bicycle.

This complete streets communications toolkit includes four components.

1.   The basics for using complete streets.

2.   Using complete streets in everyday communications.

3.   The complete streets response to a cyclist or pedestrian death or injury.

4.   Using complete streets to build coalitions.

The Basics for Using Complete Streets

The term complete streets is a description of streets that have been built for safe and convenient travel by all road users. It also describes policies that call for routinely providing for all modes when building and reconstructing streets. While the principle will most often be invoked for better walking and bicycling, complete streets should also provide safe and convenient transit access and provisions for people with disabilities. Making common cause with these users is an important element in promoting complete streets policies.

Note that complete streets is not capitalized in general use. The phrase is not proprietary and we wanted to discourage any trend toward a narrow definition of the ultimate 'Complete Street.'

A *campaign* to institute a complete streets policy can have a more formal name: Complete the Streets. Complete streets was initially coined by America Bikes in 2004 as part of the campaign to reauthorize the federal transportation law, and this campaign used the following two taglines:

•   Complete the Streets—for safer bicycling and walkable communities.

•   Complete the Streets—for safer bicycling and walking.

You can use these tags, but feel free to follow Complete the Streets with other secondary phrases. Already one organization has modified it for their campaign's name to include the health message: "Complete the Streets for Active Communities." You will want to choose one phrase and stick to it. Consistency is vital in good communications work.

The National Complete Streets Coalition, a collaborative of organizations working towards complete streets including the Thunderhead Alliance, has created some tools for those interested in advancing the complete streets cause. Many resources and a customizable PowerPoint presentation explaining the principle are available on the coalition's website www.completestreets.org

Using Complete Streets in Everyday Communications

You need to begin the complete streets transformation right away. Start by updating your existing communications. Then use it in new communications. Get your allies to start using complete streets; and have resources available for others to use.

Adjust your current communications: If you've been using the term 'routine accommodation' simply replace it with 'complete streets' in your communication materials. Look at:

- policy statements,
- brochures describing your organizational goals,
- newsletter articles, and
- website.

A few examples of how to do this follow.

Original Newsletter Headline	Recommended Change
"All new roads lead to 'routine accommodation' for bikes, peds."	"This new policy will complete the streets for bicycling and walking"

Original Text	Recommended Change
"The most efficient and least costly method to implement improvements is for the governing body to require that their transportation planners and engineers routinely improve the compatibility of bicycling and walking into every road and transit project. This will institutionalize bicycle and pedestrian planning and a "bicycle network" will emerge in short order."	"The most efficient and least costly method to implement improvements is for the governing body to require that their transportation planners and engineers design every road and transit project to be complete–safe and convenient for bicycling and walking. This will institutionalize bicycle and pedestrian planning and a "bicycle network" will emerge in short order."

Original Text	Recommended Change
"Good community design can increase opportunities for physical activity. Some examples of community design that promotes active living include:	"Good community design can increase opportunities for physical activity. Some examples of community design that promotes active living include:
• Providing routine accommodation for bicycle and pedestrian infrastructure in transportation projects."	• Completing all streets with safe accommodations for bicycling and walking.

While you may have become comfortable using 'routine accommodation,' try your best to eliminate it in all of your communication materials. It does not resonate with decision makers or the general public like complete streets does.

<u>Look for new places to use the phrase:</u> Next, you need to seek out those materials and situations where you can promulgate complete streets. Think of things like:

- letters to the editor, and

- public hearing testimony.

Here is an example:

> *"If there is inequity in the transportation system, it lies in the fact that we as Americans fail to complete our streets for safer bicycling and walking." (letter to the editor, Asbury Park Press, by John Boyle, Bicycle Coalition of Greater Philadelphia 1/22/04)*

<u>Ask your allies to use it:</u> You have allies who want you, and our bicycle and pedestrian issues, to succeed. Asking them to use complete streets in their meetings, memos and discussions is a direct opportunity and easy way that they can help. Ask allies like:

- bicycle/pedestrian planners,

- MPO officials,

- elected officials,

- smart growth advocates, and

- safety advocates.

<u>Disseminate complete streets resources:</u> You can also put some of your organization's resources to work highlighting the principle. Consider:

- adding a link on your web site to Thunderhead's National Complete the Streets Campaign web page: www.thunderheadalliance.org/completestreets.htm as well as one for the coalition: www.completestreets.org

- presenting or posting to your website the complete streets PowerPoint (with updated, local images and information),

- creating a brochure or webpage about complete streets for your communities, and

- collecting photos of complete streets and streets needing to be completed in your community.

<u>Avoiding pitfalls:</u> In your communications work, don't get bogged down trying to do the job of an engineer or planner. Stay focused on communicating the principle of complete streets. Complete streets policies are by necessity flexible and do not prescribe a single type of accommodation.

If reporters or officials try to pin you down about whether a complete streets policy will result in a specific type of facility, defer to the expertise of planners and engineers and focus on achieving the *outcome* of complete streets. Say to them, for example:

"I'm not sure what the best answer is for Smith Street, but I know the engineers and planners can come up with a solution that makes sure this important roadway is a complete street with safe provisions for people on foot and bicycle."

Be careful not to use complete streets to describe "poser" policies that leave so much wiggle room that they become meaningless, or that restrict accommodation only to roads in a bicycle or pedestrian plan. If you believe your complete streets policy is a strong policy, focus on how the policy will result in change on the ground.

A Complete the Streets Response to a Death or Injury:

Currently, every Thunderhead organization must respond to the deaths of cyclists and pedestrians in traffic crashes due to our nation's epidemic of incomplete streets. A full 13% of traffic deaths across the U.S. are bicyclists and pedestrians.[1] These deaths are often the most serious consequence of a transportation system hostile to people on foot and bicycle, but they are usually regarded by the media as individual, unavoidable tragedies in which the cyclist or pedestrian victim is blamed. Official responses often focus on education rather than fundamental change, or accept it as inevitable (e.g. "The driver just didn't see him/her. It is a tragedy."). It is the Thunderhead leaders' role to point out the more fundamental problems that likely contributed to the death or injury.

The complete streets concept can help you respond to such deaths in a way that educates people about the deficiencies in our transportation network. However, be sure that the facts fit this issue. A visit to the crash scene may be necessary to evaluate the street to see if an incomplete street may have been a factor in the crash. In addition, avoid a fight over whether the incomplete street 'caused' the death. Such a determination is the responsibility of the police and justice system. Simply state that the road was inadequate and point out this could have been one factor in the crash.

Once you have determined that an incomplete street could have been a factor in the death:

- Determine exactly what is missing from the street where the crash occurred. Does the city/county/state have any future plans to improve the road?

- Decide whether to point out a specific problem or failure or if your message is more general education on why complete streets are necessary. This will be determined largely by your relationship with local governments.

- Write and distribute a complete streets news release—quickly. Coverage of such deaths will likely be short-lived. See the sample news release below.

- Be sure to call the reporters who are covering the story to give them your perspective and to encourage follow up stories on conditions for bicycling or walking. Again, respond quickly. Reporters won't be interested a week after the crash.

---

1.   Fatality Analysis Reporting System

- Write a letter to the editor. This is quick and easy and is a good fallback if you cannot get news coverage. See the sample letter below.

Sample News Release:

This news release is designed for general education; you can easily modify it to call for specific street improvements.

For Immediate Release   **For more information, contact:**
[date]   **[name, phone]**

Incomplete Street May Have Contributed to Cyclist' Death
[name of Thunderhead organization] calls for action

The death of cyclist/pedestrian [name] on [date] occurred on a street that is not designed for safe cycling or walking, according to [Thunderhead org.]

"[name] was riding on a street that is incomplete—it is designed without room for safe cycling," said [org leader]. "To prevent future deaths, our [local government] needs to start creating complete streets that are safe for people traveling by car, foot, or bicycle." [see additional sample quotes below]

While the police will determine who was at fault in the crash, the fact that no provision was made for motorists and cyclists to share the road may well have been a factor. [include details here about what the road is missing]

[Thunderhead org] has been urging [local government] to institute a complete streets policy, so that every road will be made safe for bicycling and walking. Complete streets can be created by building sidewalks or bike lanes, widening curb lanes, improving shoulders and intersections, or by installing traffic calming devices to slow traffic.
OR
Streets such as [road name] can be completed by building sidewalks or bike lanes, widening curb lanes, improving shoulders and intersections, or by installing traffic calming devices to slow traffic.

"Each complete street may look different. We are only asking that when engineers build or reconstruct a road, they take travel by foot and bicycle into account," says [org leader].

"I support creating complete streets to avoid future tragic deaths and to give residents of [jurisdiction] safer places to bicycle and walk," says [local political leader.]

[jurisdiction] has a bicycle plan, but it only covers some streets, and [road name] is not one of them. A complete streets policy would ensure that eventually every road would make provision for people on foot and bicycle.

OR

[jurisdiction] has a bicycle plan, but this street has not yet been upgraded in accordance with the plan. "The fact that a cyclist has lost their life/been critically injured demonstrates the urgent need for these improvements."

[one sentence about advocacy group]

For further information, contact:

A couple of additional sample first quotes for news release are:

- "This death occurred on a street that has narrow, high speed lanes and no sidewalks. We call this an incomplete street—because it only provides for safe travel via automobiles and does not provide for travel on foot and bicycle," says [organization leader.]
- "[name] was riding through an intersection that does not provide for safe travel by foot or bicycle," says [organization leader]. "This high-speed road does not have enough space or proper signals for non-motorized users. The [local government] needs to do more to create safe places to walk or bicycle."

Sample Letter To The Editor

Dear Editor,

The death/serious injury of [name] while riding a bicycle/walking on [road] is a heart-rending demonstration of why we need to do more to make our streets safer for bicycling and walking. [Road] is an incomplete street—it does not have provisions for safe travel via bicycle or foot.

[our organization] is urging the city/county/state to start to build complete streets—roads that are safer for all travelers. Streets such as [road name] can be completed by building sidewalks or bike lanes, widening curb lanes, improving shoulders and intersections, or by installing traffic calming devices to slow traffic. Each complete street may look different. But when engineers build or reconstruct a road, they must take travel by foot and bicycle into account.

Sincerely,

[Executive Director,
Bicycle/Pedestrian Advocacy Organization]

Using Complete Streets to Build Coalitions

Complete the Streets campaigns are a potential tool for creating diverse, powerful coalitions. The complete streets concept goes beyond the narrow focus of providing bike lanes and sidewalks. It is about routinely ensuring the public right-of-way serves everyone. That means it has the potential for broad appeal.

National momentum is building: The push for complete streets is taking hold in many sectors from smart growth to the disabled community, from developers to progressive agency representatives. Many of these progressive organizations have been brought together to form the National Complete Streets Coalition. Thunderhead's National Complete the Streets Campaign is offering our state and local campaign efforts as part of this coalition.

Listed below are the current active groups in the National Complete Streets Coalition. More groups are looking into joining the coalition and taking part in its efforts.

- AARP
- American Society of Landscape Architects
- American Planning Association
- America Walks
- Association of Pedestrian and Bicycle Professionals
- Institute of Transportation Engineers
- League of American Bicyclists
- National Parks Conservation Association
- Smart Growth America (closely allied with the 1,000 Friends-type groups)
- Surface Transportation Policy Project
- Thunderhead Alliance
- US Access Board

The Coalition is focusing its efforts on spreading the word about the benefits of complete streets. Please visit the Coalition's web site: www.completestreets.org. As part of the Coalition, the Thunderhead Alliance is developing our strategy to help you connect with local representatives and allies of these organizations and we can help you with these connections. If you already work with local representatives of these organizations, you can let them know that their national organization is getting behind complete streets.

As you can see from the list above, the concept of complete streets is also being adopted by engineers and planners who have a long-standing interest in the closely-related campaigns for Main Street revitalization, context-sensitive design, and multi-modal planning. For example, the Transportation Research Board's annual meeting (the mainstream transportation professionals conference) began featuring well-attended complete streets sessions and workshops in 2005, with sessions co-sponsored by a mainstream committee normally associated with highway design.

Context-sensitive design, also known as context-sensitive solutions, is an internal movement to better integrate road projects with their surrounding community, and often includes discussion of bicycle and pedestrian facilities. This project-by-project approach, which usually is launched for a major project and involves extensive stakeholder participation, compliments the complete streets policy approach so long as complete streets are part of the discussions.

The USA EPA, the Institute of Transportation Engineers, and the Congress for the New Urbanism (CNU) are putting out an "Urban Arterials Guide" which tackles the full complexity of designing roads that work for all users as well as the destinations adjacent to the road. As Thunderhead leaders you can help bring these new resources into your communities.

Another issue to watch is the intersection between the Americans with Disabilities Act and complete streets. The FHWA has issued a draft of a new policy directive that directs transportation agencies to meet accommodation requirements by considering the entire right-of-way for each road improvement project. For example, a roadway reconstruction project would have to consider pedestrian access. This could be a potentially powerful tool in pushing for complete streets.

Building support for your campaign: In building your own local campaign, you can approach many potential supporters including others who use your communities' roads, public health groups, smart growth organizations, community development organizations, progressive planners and engineers, and others.

*Others who use the roads: P*eople with disabilities, representatives of older people and children (AARP, PTA), transit users and transit agencies, even truck drivers, motorcyclists, motor-vehicle drivers—they all benefit from having safer places for everyone to travel. For example, you may be able to work for complete streets on a micro-level by making common cause with neighborhood groups dealing with traffic. Traffic calming should be considered a piece of the complete streets picture. Or you may already have a relationship with your transit agency. Talk to them about how they now interact with your DOT or public works department.

Public health groups and staff: These groups want to increase walking and bicycling to improve health and may lend important credibility and support. These groups may include the state chronic disease and injury prevention staff, your local public health agency, the American Heart Association, or the American Lung Association. The local parks department may also be involved in the active living movement. More and more parks are helping promote physical activity through bicycling and walking, both inside and outside park borders.

Smart growth groups: This may include a local smart growth advocacy group, a smarter growth business coalition, and environmental groups. They most likely are already familiar with creating walkable communities, and may be involved in promotion of traffic calming, so complete streets should be an easy sell.

Community development groups: Many towns have redesigned their main streets to be more pedestrian friendly as part of plans to revitalize their community. The Chamber of Commerce or Community Development Corporations may be receptive to taking the idea beyond main street and pedestrian-only issues.

Progressive planners and engineers: You know who they are. You can tell them that the American Planning Association and the Institute of Transportation Engineers are supportive of the concept.

Get Creative!: The complete streets concept also has broad enough appeal for educating or enlisting groups as diverse as the League of Women Voters and the Rotary Club.

<u>Ways to enlist campaign supporters:</u> Advocates can begin a Complete the Streets campaign by interviewing people from some of these groups. Would a complete streets policy help them meet their goals? What problems would complete streets address for their constituency? What would be essential for them to support a Complete the Streets campaign? This could begin with a presentation using the basic Complete the Streets PowerPoint available on the complete streets website. This is a downloadable version which can be customized.

Then, have a conversation with them. Ask if a complete streets policy would help them meet their goals. What problems would complete streets address for their constituency? What would be essential for them to support a Complete the Streets campaign? You will probably learn quite a bit through this process, and will be building a valuable long-term partnership.

Even if you are not ready to launch a full-blown Complete the Streets campaign, you can be sure these discussions continue by asking them to sign a petition, or write a letter of support, or endorse a proposed resolution. This can be important early groundwork for a campaign.

Another way to build support, particularly among community groups and decision makers, is to bring in an outside speaker or expert. A number of consultants and groups conduct 'walkable community workshops' and, if they are involved with creating complete streets, can easily expand these for complete streets. They will come in to introduce these concepts and help communities solve transportation problems. The Pedestrian and Bicycle Information Center's Pedestrian Audit Course and Safe Routes to School Course as well as the National Center for Bicycling and Walking's Walkable Communities Workshops are three such resources. While these programs generally work through problems on specific streets or corridors or school zones, they can be used to support the more general concept of providing complete streets, every time.

If you know that you won't be able to launch a campaign for a complete streets policy any time in the near future, consider submitting short articles about the complete streets concept to the newsletters of your target organizations. Just getting more people using the phrase is an important early goal.

Building a strong and diverse coalition will be an essential element of any full-blown Complete the Streets campaign. These suggestions are only the beginning of that process. Be sure to share your innovative ideas with us as your campaign progresses!

A Note From Sue Knaup, Thunderhead's Executive Director, on Next Steps:

Now it's time for you to take what you have learned from this Guide, plug in your own talents and innovations, and bring your unique campaign to your communities. As the leader of a Thunderhead organization, you are one of the few gifted people in this world able to effectively turn rapid-fire assaults into clear paths, see opportunities where others see only barriers, and juggle the needs of your staff, members, allies and future allies with a smile that stems from your indelible vision of your communities as havens for bicycling and walking. Thunderhead invested in this Guide to help you succeed with your campaigns because we know you are the heart and muscle of the bicycle and pedestrian advocacy movement. We look forward to assisting you as you move ahead.

Our updates of this Guide will depend on your experience with your campaigns. You are pioneers of a new territory that is sure to shift the tide of our nation's transportation expectations. In order to help you and new leaders who take on complete streets campaigns we need to learn from you and what you discover in your campaigns.

We published our first edition of this Guide in March 2005. One year later we proudly bring to press this first update. We expect to publish further updates as our National Complete the Streets Campaign moves forward.

Please let us know as soon as you have launched your complete streets campaign so that we can connect you to opportunities that we discover through our networks. Then, as you move ahead, please send your insights and suggestions on campaign development, policy language, policy implementation and innovative advocacy approaches to me. Thank you for participating and for your commitment to our movement.

Sincerely,

Sue Knaup
Executive Director
sue@thunderheadalliance.org

APPENDIX A

Campaign Examples

Example 1: Local Option Sales Tax (for Bike/Ped and Safe Routes to School)

Title: Measure A: Transportation Sales Tax

Thunderhead
organization: Marin County Bicycle Coalition

Location: Marin County, California

Level: Local

Type of Campaign: Legislation

Description: A half-cent sales tax increase that will generate approximately $331 million over the next 20 years dedicated to local transportation projects, including $36 million for Safe Routes to School and a complete streets policy. All projects will consider all users, including transit, bicyclists and pedestrians.

Adoption date: 11/02/04

Policy online: www.marinbike.org/Campaigns/Infrastructure/MeasureAPlan.pdf

Thunderhead leader
time involved: 6+ years. Staff attended <u>every</u> meeting of the Transportation Authority of Marin. MCBC committed to working on the passage of the transportation tax the day after its defeat in 1998.

Organization direct
benefit: 1) Bragging rights to membership by telling them MCBC helped shape Measure A, the most powerful transportation tax that will come along in 20 years; 2) MCBC currently operates Safe Routes to School program under a $240,000 contract with the city, providing funding for 3.5 FTE staff.

Indirect benefit:	1) Instead of being on the outside, MCBC is on the inside and considered a team player by city agencies; 2) Showed elected officials and city agencies that bike people had the power to get voters mobilized and get things done.
Issue focus:	Traffic is bad and getting worse. The public wanted more places to walk and bike safely. They were ready to participate and hear other solutions, especially since the transportation tax failed on three other attempts in 1980, 1990 and 1998. The public was ready to participate and give their input into how to fund transportation alternatives while considering the needs of bicyclists and pedestrians.
Campaign goals:	Secure voter approval of a half-cent sales tax increase that would generate roughly $331 million over 20 years, in four key strategies: school access; infrastructure; transit and HOV and bikeway. The overall goal was to develop a plan to increase mobility and reduce traffic congestion.
Strengths/ Weaknesses:	MCBC Board buy-in was strong. Marin County Supervisor Steve Kinsey, who chaired the Transportation Authority of Marin, took over as chairman in 1998 and had an amazing vision of how to integrate the various modes of transportation. The Transportation Authority hired a top-notch transportation consultant who worked on the passage of other successful tax ordinances and was able to steer the campaign in a direction that worked.
Allies:	It was a first for environmentalists and businesses to work together. Once the transportation sales expenditure plan was created, more than 100+ organizations joined as allies to pass the measure with 71% of the vote. Supporters included the Association of Realtors, Marin Commission on Aging, Sierra Club, Greenbelt Alliance, Builders Association, and Transportation Alternatives for Marin.
Opponents:	Tax payer union.
Strategy:	MCBC had a two-part strategy. First, get bicycle and pedestrian elements included in each of the four strategies designated in the transportation sales tax expenditure plan. Meetings with Marin County supervisor allowed MCBC to hear what would work and what would win if they were going to fit what they wanted into the four expenditure categories previously established. MCBC showed how they were supporting the Measure's overall goals by identifying how MCBC goals related to theirs. The second strategy was helping get the transportation sales tax plan approved by two-thirds of the voters.

Targets

Decision makers: Transportation Authority of Marin served as the decision making body for the elements in the plan and included representatives from the County Board of Supervisors and each of the 11 cities and towns.

Public audiences: Parents of Safe Routes to School students and education officials

Communication methods: Organized letters to the editor efforts among supporters and kept members informed through the newsletter and website. Trained parents to speak on congestion management, recommending bicycling and walking as alternatives to being driven in motor vehicles. MCBC used the voter polls to show the existing public support for Safe Routes to School. They provided easy to use sound byte quotes for the media and stayed consistent on their message "bicyclists and pedestrians are part of the solution."

Tactics: MCBC staff attended every meeting of Transportation Authority of Marin for six years to serve as the voice for bicyclists and pedestrians and to get bike/ped elements included in the expenditure plan. They created a Position Paper outlining their goals and the exact amount of funding they desired in each of the four strategies presented in the plan. They showed how bicycling and walking improvements would improve the mobility in their community and provided sample language. The Paper was heavily circulated to the Transportation Authority and the five Citizens Advisory Committees. MCBC organized rallies of school children and people with disabilities to garner positive media attention. They also developed e-mail alerts, downloadable posters and held phone banking out of their office. Election day morning found forty volunteers organized by MCBC positioned at strategic freeway entrances holding signs saying "Yes to Measure A."

Resource
Management:
Budget: Measure A was included among MCBC's list of projects in their membership solicitation and fundraising materials.

Volunteers: Coordinated more than 100 volunteers during the six-year campaign, especially to hold signs on election morning.

Keys to campaign success: 1) consistent message for 6 years; 2) used Position Paper to show why voters would support walking and bicycling; 3) by demonstrating strong grassroots organizing to get people to show up left impression as strong team player and earned a seat at the table.

Pros/Cons of
Campaign: None

Things to do
Different: If other communities embark on a transportation sales tax, realize that despite what could be a long-term haul, you have to know the biggest payoff comes at the end and is only possible if you stick with it and put forth your vision. By attending every meeting, MCBC staff was better able to assess and know when to call out the troops.

Other comments: After the four category expenditure amounts were finalized, the High Occupancy Vehicle lanes category received an extra $10 million in funding from an outside source. The Commissioners debated about changing the percentage from $25 million to $15 million and allocating the extra money to a different category. Because Executive Director Deb Hubsmith was at the meeting, she advised them during the public comment period to keep the $10 million in the original category and to expand the category's definition to include bicycles. Deb suggested that the money could be used for the completion of the much-needed 2-mile section of multi-use path on the North-South Greenway. What could happen in 30 seconds? The members of the Commission took about 30 seconds to deliberate over Deb's suggestion and instead of shifting the money to another category, they kept the $10 million in the original category and expanded the definition to include bicycles. It pays to show up.

Example 2: Statewide Legislation (Safe Routes to Schools)

Title: Texas Safe Routes to School

Thunderhead
organization: Texas Bicycle Coalition

Location: Texas

Level: Statewide

Type of Campaign: Legislation

Description: <u>House Bill 2204</u>, 77th Texas Legislature, directed the Texas Department of Transportation (Texas DOT) to establish the Safe Routes to School Program.

The grassroots campaign after the bill's passage focused on generating high demand for the first Call for Project applications. TBC had less than 4 months from the time the announcement was made in August 2002 to demonstrate to the Texas DOT Commissioners the popularity and demand of the new program.

Adoption date: June 15, 2001(signed by Governor)

Policy online: http://www.capitol.state.tx.us/cgi-bin/db2www/tlo/billhist/actions.d2w/report?LEG=77&SESS=R&CHAMBER=H&BILLTYPE=B&BILLSUFFIX=02204
                       www.saferoutestexas.org

Thunderhead leader
time involved: 3.5+ years; 2 years to pass legislation; 1 year to adopt the project selection rules; and 6 months to promote the first Call for Projects.

Organization direct
benefit: 1) program's popularity generated $45 million in requests for a $3 million program call; 2) DOT increased funding to $5 million after seeing the demand in the form of more than 300 project applications in less than 4 months; 3) In October 2004, TBC was awarded a three-year federal grant for $1.5 million to administer a Safe Routes to School education and encouragement program in 300 schools and 27 cities in Northwest Texas.

Indirect benefit: 1) Shifted the debate with the Texas DOT about creating good bike and ped facilities; 2) witnessed DOT's attitude change from resistant to enthusiastically embracing the program over 24-month period; 3) built stronger relationships and increased credibility with Texas DOT and state and federal legislators; 4) Positioned TBC to be a leader within Texas on Safe Routes to School.

Issue Focus:	Schoolchildren are not getting enough healthy exercise and traffic congestion around schools is burgeoning; safe biking and pedestrian facilities in and around schools which encourage self-transportation will relieve both situations.
Campaign goals:	Get legislation passed, then: 1) Create a citizen's advisory committee for the project application selection; 2) create high demand through promotion and encouragement for communities to partner with their school districts and submit an application within the 4-month application period; 3) Encourage 250+ bike shops to play a prominent role in promoting the Call for Projects within their community.
Strengths/ Weaknesses:	1) TBC was well prepared before the Texas Legislative session started; hired a professional lobbyist, along with a volunteer campaign consultant and volunteer campaign manager; 2) difficult for politicians to be against safe children; 3) completed major overhaul of website before the legislative session
Allies:	High profile groups such as Texas Medical Association, Texas Association of Parents and Teachers, Texas Hospital Association, Texas Association of Health, Physical Education, Recreation and Dance.
Opponents:	Initial resistance by some legislators and Texas DOT
Targets: Decision makers:	State legislators, Texas DOT Commissioners
Public audiences:	PTA, school superintendents, teachers, mayors and city managers
Communication	During legislative session, website generated 350+ new members; during Call for Project period, 300+ newspapers printed articles over four months; email action alerts; great public relations tool by using bike shop's customer list and crossed with enhanced voter registration lists to customize all mailings.
Tactics:	Use enhanced voter list to find bicyclists and "Super Voters", targeted letters to transportation commissioners; identify constituents to send letters to committee members before public hearings; e-mail was a "new" technology in reaching legislators in 2001, generating thousands of responses which took elected officials by surprise; kept supporters informed with daily posting of new activities on website; avoided list fatigue by only sending members action alerts for their specific legislative districts; give legislators a new impression of range of constituents by having attorneys, doctors, engineers, and even a 12-year old testify, who were all bicyclists, testify at committee hearings. TBC did not overplay their presence in the committee hearings by having too many people testify

Resource Management:	
Budget:	Fundraising plan included bike shops ($37K); individual donors through special appeal letter and website ($30K) plus 350+ new members; and Bikes Belong ($10k) grant supplied creation of new website dedicated to Safe Routes to School and promotional materials for Call for Projects announcement.
Volunteers	More than 100 cyclists attended the 1st Cyclists in Suits Bike Lobby Day at the Texas Capitol; list serve managers from different bike clubs and websites would forward action alerts to their members.
Pros/Cons of Campaign:	Pros—1) increased credibility and enlarged pool of organizational and political allies and potential allies; 2) broadened the constituency for communities realizing they could have bicycling and walking as part of their plans; 3) involved bike shops as way to reach into communities Cons—1) did not anticipate one year of unfunded staff time during the rulemaking and adoption process; 2) became identified as clearinghouse for Safe Routes to School in Texas and spent a lot of time answering questions about the application process that was not funded; 3) risk of staff burnout
Things to do Different:	1) Must secure additional funding sources after any type of legislation victory for promotion and administrative rules adoption process; 2) Prepare bike shop owners with specific instructions for "enhancing" their customer lists with voter registration lists <u>before</u> the campaign so you have everything in-house and are ready to respond with targeted mailings in the moment of urgency.
Other comments:	1) Positive hype about program's gaining popularity enabled the bill's sponsor to announce $3 million in funding almost 8 months after the bill passed; 2) Able to win over the senator of a key committee by sending 6,000 "friendly" pieces of mail to constituents in the senator's district, acknowledging her solid support on bicycling issues and informing the legislator's staff.

Example 3: Local Policy (Complete Streets)

Title:	MORPC Bicycle and Pedestrian Planning Policy: Routine Accommodations 2004
Thunderhead organization	Central Ohio Bicycle Advocacy Coalition
Location:	Mid-Ohio Regional Planning Commission (Metropolitan Planning Organization for Central Ohio)
Type of Campaign:	Resolution by MPO with detailed policy
Description:	Required to accommodate bicycles and pedestrians in the planning and design of all proposed transportation projects using MORPC-attributable federal funds.
Adoption date:	07/22/04
Policy online at:	www.morpc.org; www.cobac.org
Thunderhead leader time involved:	15 months and 200 hours. Between April 2003 when COBAC lodged objections to MORPC TIP and adoption of MORPC routine accommodation policy on July 22, 2004, approximately 200 hours were spent on advocating for this policy. Campaign materials to share: More than 15 attachments available by request through Thunderhead Alliance
Organization direct benefit:	1) changed local policy so all transportation agencies in the state adopt the policy; 2) established organizational credibility; 3) natural transition to switch from MORPC success at the local level to pursue the state department of transportation to adopt the policy. Indirectly, the organization's success of passing this policy is helping to establish and build the need for this type of advocacy organization
Issue Focus:	A review of the MORPC Transportation Improvement Plan revealed almost none of the projects approved included accommodations for bicycling and walking. In April 2003, COBAC objected to MORPC's TIP and to granting federal funding to projects due to failure to comply with federal law requiring "due consideration".
Campaign goals:	Encourage MORPC to adopt a complete streets policy. This subsequent success led COBAC to prioritize other goals and apply pressure to the Ohio Department of Transportation. None of the 1,400 projects listed in the Ohio DOT Statewide Transportation Improvement Plan included accommodations for bicycling or walking.

Strengths/ Weaknesses:	Volunteer campaign with no paid staff.
Allies:	MORPC staff
Opponents:	No organized opposition, except Franklin and Columbus County Engineers objected to language that could have been interpreted as setting a minimum percentage to be spent on accommodating bicycling and walking in each project.
Strategy:	COBAC submitted a letter to MORPC in April 2003 demanding that all federal funds be withheld from all projects listed in the TIP that did not include bicycling and walking as required by federal transportation law and policy. In May 2003, COBAC sent a letter to the Federal Highway Administration with objections regarding Ohio DOT's 2004-07 STIP, copying Ohio DOT.
Targets Primary:	MORPC Commissioners, FHWA Division Office and Ohio DOT director
Secondary:	Ohio Governor
Public audiences:	Bicycle retailers and clubs, alternative transportation organization and Association of Railroad Passengers
Communication Methods:	Used rhetorical arguments; public health information and sample text from other jurisdictions in all correspondence and testimony.
Tactics:	Attended official meetings; submitted comments; encouraged testimony at public hearings; worked the political process; and helped write and revise language. Resource Mgt. Budget: $0; volunteer campaign with no fundraising plan
Volunteers:	COBAC president served as primary volunteer; enlisted support from 8 organizations in letter to Federal Highway Administration objecting to Ohio's Draft FY 2004-2007 Statewide Transportation Improvement Program.
Keys to campaign success:	1) Fortunate enough to have forward looking MPO that really did want the policy; 2) rivalry among neighboring states and MPO's; 3) COBAC president served as a member of the citizen advisory committee. This helped him stay connected with MPO plans, process and staff.
Pros/Cons of Campaign:	1) Cemented relationship with MPO and the jurisdictions; 2) COBAC is not perceived as narrow interest group and the process got COBAC working

with MPO; 3) Downside from MORPC's point of view was the aggressive tone or "demand" of the letter, indicating that the tone of the letter was not necessary.

Things to do
Different: 1) Use less aggressive language in the first letter to MORPC if advocacy organization had known MORPC better; 2) include more pedestrian and disabled organizations as coalition partners but they were not interested.

Other comments: Keys to policy success: 1) supportive, sympathetic staff at MPO; 2) adoption of policy at rival MPO in northeast Ohio in fall of 2003 challenged leadership position of MORPC; 3) serious threat to federal funding for local transportation projects if they did not adopt routine accommodation policy. COBAC has periodic meetings with bike/ped planner to check on how the new policy is working with MORPC. COBAC wants to ensure new policy has measurable results and has suggested using the PennDOT Bicycle/ Pedestrian Facilities checklist of July 2001.

APPENDIX B

Blueprint Worksheet

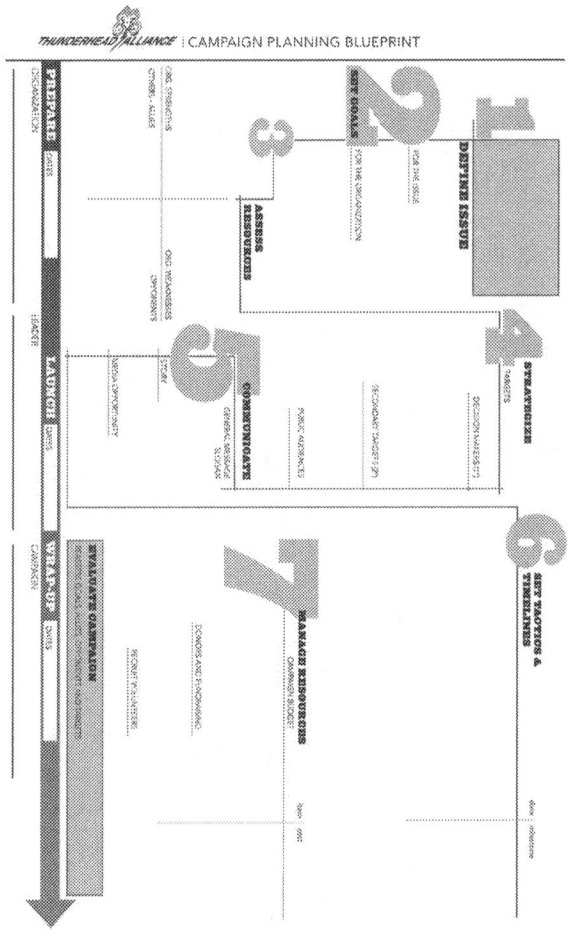

Please see the member side of the Thunderhead website (www.thunderheadalliance.org) for a full, legal size, version of this blueprint.

APPENDIX C

Complete The Streets Survey Form

This survey was distributed to Thunderhead organization leaders and bicycle-pedestrian planners.

Thank you for taking the time to answer the Thunderhead Alliance survey on complete streets policies. We are using this information to create an inventory of policies now in place as well as of active campaigns to institute complete streets policies.

Please answer the following questions to the best of your ability: We realize that you may not have the answers to every question. To answer the survey without disturbing its layout, hit the 'Insert' key on your keyboard or double-click the OVR button in the status bar at the bottom of the Word screen.

You may return the survey electronically by emailing it to barbara@bmccann.net. Questions or comments? Contact Barbara McCann at 202-641-1163 or barbara@bmccann.net.

Your Name:
Title:

Phone:
Email address:

What is a complete streets policy?	What is not a complete streets policy?
Complete street policies call for creating safe and convenient bicycle and pedestrian accommodation on every road built or reconstructed (sometimes called 'routine accommodation'). These facilities can be quite varied, ranging from separate paths to sidewalks and bike lanes to wide shoulders or wide curb lanes, but a justification is required if no bicycling and/or walking improvements are made.	Policies that: • limit consideration of accommodation to roads in a bike or pedestrian plan, • ask for some justification of need before a bicycle or pedestrian accommodation facility is included, and/or • simply encourage consideration without any requirement.

We do want to include policies which appear to require complete streets, but which have not succeeded in doing so because of implementation failures.

Questions:

What is the name of the policy?

What jurisdiction does the policy cover?

What is the origin of the policy?
____ State legislation
____ City/county council resolution/ordinance
____ Internal DOT guidance or directive
____ Integration with street design manual
____ Other: _____

When was it adopted? (For advocates working toward new policies, see below)

Where can the complete original language are found on the Internet?

Was there a press release on its adoption/implementation and where can a copy be located/obtained?

If not available on the Internet, can you give a one-sentence summary of the policy?

What are the exceptions allowed in the policy?
_____ Excessive cost
_____ Absence of need
         Bridges
_____ Insufficient right of way
_____ Conflict with local plans
_____ At the discretion of a top official

____ Other: _____

Must the exceptions be formally justified?
____ Yes____ No

Is the policy being successfully implemented?
____ Yes ____ No

What issues are hampering implementation?

What are the financial considerations surrounding the policy? Does it include dedicated funding, either for facilities or for administration?

Are there any quantifiable outcomes being tracked at this point, such as number of streets 'completed'; an increase in biking/walking; or any other statistics?
____ Yes____ No

Can you provide some of these statistics here?

Were advocates involved in getting the policy adopted?
_____ Yes____ No
If yes, which organizations and/or individuals were involved?

Have advocates been involved in moving toward implementation?

Has there been any opposition to the policy?
_____ Yes____ No
If yes, can you tell us what opponents have said and who the opponents represent?

Who else should be contacted for further information (please provide a phone number or e-mail address)?

Is there anything else you would like to add?

Additional Questions for Advocates

If you are an advocate working toward or have a complete streets policy now, please answer these questions according to what has been done to date.

Is your policy in place, or in process?
____ In place _____ In process

Did you originate the movement for a complete streets policy, or support an effort begun by other players (elected officials, transportation or planning department officials)?
____ Originated with advocates ____ Originated by others

Who have been your biggest allies and supporters in seeking and implementing complete streets policy?

How much time have you or your organization spent in total advocating for adoption of the policy?

If policy is in place, how long did it take from the first introduction of the idea to implementation?

What activities have you engaged in while working to achieve the policy?
____ Attended official meetings and submitted comments
____ Arranged meetings with officials
____ Circulated petitions
____ Engaged in a public media campaign
____ Encouraged testimony by members at public hearings
____ Worked the political process
____ Helped write and revise language
____ Other: _____

How much has this work cost your organization?

What information helped with your advocacy?
____ Local biking/walking statistics (use, crashes, etc)
____ Rhetorical arguments for balanced transportation
____ Fiscal arguments
____ Public health information
____ Technical information on feasibility
____ Sample text and examples from other entities/jurisdictions
____ Other:

What do you believe have been the top three keys to your complete streets success?

What three things would you do differently if you were starting from scratch on your complete streets effort?

Thanks again for completing this survey! Please email it back to barbara@bmccann.net.

APPENDIX D

Policies Surveyed

Policy	Level	Type	Adopted	Description	Original source
FHWA policy		policy guidance	02/28/00	Original FHWA guidance based on language in TEA-21.	www.fhwa.dot.gov/environment/bikeped/design.htm#d4
California DDOT Deputy Directive 64	state	internal policy (Deputy Directive 64)	03/26/01	"The Department fully considers the needs of non-motorized travelers (including pedestrians, bicyclists and persons with disabilities) in all programming, planning, maintenance, construction, operations and project development activities and products. Adopts best practices from USDOT policy statement.	www.dot.ca.gov/hq/tpp/offices/bike/DD64.pdf Note at: www.calbike.org/acr211.asp you can see the state legislature's August 2002 resolution urging local jurisdictions to adhere to DD-64 and the FHWA guidance document.
Sacramento County, California routine accommoda-tion sales tax initiative	county & all cities in county	tax ordinance, 30-year sales tax	11/02/04	One sentence requires routine accommodation of bicyclists and pedestrians in all projects funded by half-cent sales tax.	www.sta.sacramento.ca.us/pdf/OrdSTA-04-01.pdf
San Diego, California City Street Design Manual	city	manual	11/25/02	Basically, every street is required to have bicycle and pedestrian accommodation.	www.sandiego.gov/planning/pdf/intro.pdf

Santa Barbara, California Circulation Element, General Plan	city	plan, general	09/01/98	Policies direct sidewalks, bike lanes, improved roads, consider all modes when doing project; "achieve equality of convenience and choice among modes."	Please find the Circulation Element link at: www. santabarbaraca.gov/Government/Departments/PW/Transportation+Planning+and+Alternative+Transportation.htm
Sacramento, California Pedestrian Friendly Street Standards	city	resolution of city council amending general plan	02/24/04	Street design manual that integrates bike/ped: Eliminate rolled curb; Include separated sidewalk on all streets; Reduce widths of collector and arterial streets; Reduce travel lane widths on arterial streets; Add bike lanes to all new collector streets.	www.pwsacramento.com/traffic/streetrevisions.html
San Diego County, CA tax reauthorization	county	tax ordinance, reauthorization of county transportation tax	11/2/04	"All new projects, or major reconstruction projects, funded by revenues provided under this Ordinance shall accommodate travel by pedestrians and bicyclists, except where pedestrians and bicyclists are prohibited by law from using a given facility or where the costs of including bikeways and walkways would be excessively disproportionate to the need or probable use. Such facilities for pedestrian and bicycle use shall be designed to the best currently available standards and guidelines."	www.sandag.org/index.asp?projectid=255&fuseaction=projects.detail See Section 4(D)(3). D
Boulder, Colorado Multimodal Corridors & Transportation Network Plans	city	Plan	01/01/96	Designated Multi-Modal Corridors are getting extra investments for auto, bike, ped & bus; Transportation Network Plans create multi-modal plans within specific geographic areas.	www3.ci.boulder.co.us/publicworks/depts/transportation/master_plan_new/multimodal/multimodal.htm

Florida Bicycle & Pedestrian Ways statute	state	legislation	1984	"Bicycle and pedestrian ways shall be given full consideration in the planning and development of transportation facilities, including the incorporation of such ways into state, regional, and local transportation plans and programs. Bicycle and pedestrian ways shall be established in conjunction with the construction, reconstruction, or other change of any state transportation facility, and special emphasis shall be given to projects in or within 1 mile of an urban area."	www.flsenate.gov/Statutes/ index.cfm?App_mode=Display_ Stat- ute&Search_String=&URL=Ch0 335/SEC065.HTM&Title=- >2003->Ch0335->Sec- tion%20065 For implementing FDOT policy, see section 8.1 of the Plans Preparation Manual, www.dot.state.fl.us/rddesign/ PPM%20Manual/2004/ Volume%201/V1Chap08.pdf
Illinois Bureau of Design & Environment, Bicycle & Ped Accommoda-tions	state h'ways	internal policy; DOT directive	09/01/95	If specific needs "warrants" are met, then curbed urban roads should include (typically) 13' outside lanes or (rarely) bike lanes, and rural roads should have paved shoulders of width depending on the situation.	www.dot.state.il.us/desenv/ BDE%20Manual/BDE/pdf/ chap17.pdf
DuPage County, Illinois Healthy Roads Initiative	county	internal directive	03/24/04	Construct a sidewalk or bicycle path where right-of-way is available; Ensure that the new construction project is safe for both the user and the community; Ensure that the new construction project adds a lasting value to both motorized and non-motorized users; couple of aesthetic concerns.	www.dupageco. org/pressDe- tail.cfm?doc_id=1352

Kentucky Pedestrian and Bicycle Travel Policy	state	internal policy	07/16/02	"The Kentucky Transportation Cabinet (KYTC) will consider the incorporation of pedestrian facilities on all new or reconstructed state-maintained roadways in existing and planned urban and suburban areas." "The Kentucky Transportation Cabinet (KYTC) will consider the accommodation of bicycles on all new or reconstructed state-maintained roadways. KYTC will also consider accommodating bicycle transportation when planning the resurfacing of roadways, including shoulders."	www.kytc.state.ky. us/Multimodal/pdf/ Task%20Force%20FINAL%20J une%2018_02%20policy%20re c%20to%20Sec%20Codell.PDF
St. Joseph, Missouri bike-ped plan	MPO	plan	07/01/01	"Bicycle and pedestrian ways shall be established in new construction and reconstruction projects throughout the metropolitan area, unless one or more of three conditions are met."	www.ci.st-joseph.mo. us/publicworks/bpmaster-plan.asp
Columbia, Missouri Model Street Standards	city	ordinance, city council	06/07/04	Subdivision ordinance: new development will include: residential streets 28' wide (instead of 32'), residential sidewalks 5' wide (instead of 4'), major collectors and arterials with 8' or 10' multi-use "pedways" and 6' striped bike lanes or wide shared-use travel lanes . These standards will be applied when streets are rebuilt, whenever possible.	www.gocolumbiamo. com/Council/Bills/2004/ apr5bills/B92-04.html

North Caro-lina DOT Bicycle Policy	state maintained roads; there no county roads in NC	resolution, State DOT	1978 and revised 1991	"...bicycling and walking accommodations shall be a routine part of the North Carolina Department of Transportation's planning, design, construction, and operations activities"	www.ncdot.org/transit/bicycle/laws/laws_resolution.html
Mid-Ohio Regional Planning Commission Bicycle and Pedestrian Planning Policy: Routine Accommoda-tions 2004	MPO	resolution of MPO with detailed policy	07/22/04	Project sponsors are required to accommodate bicycles and pedestrians in the planning and design of all proposed transportation projects using MORPC-attributable federal funds. Sponsors using local, state, or other federal funds are encouraged to accommodate bicycles and pedestrians in the planning and design of all proposed transportation projects.	www.morpc.org/web/departments/transportation/bikeped/T-15-04_Att_5-Rev_Routine_Accommodation_v2.pdf
Northeast Ohio Area-wide Coord. Agency Bike-Ped Planning Policies	MPO	internal policy	09/01/03	"Bicycle and pedestrian ways shall be established in new construction and reconstruction of road and bridge projects unless one or more of four conditions are met."	www.noaca.org/RTIP%202003.pdf page 20 (or page 15 of document)
Oregon Bicycle and Pedestrian Program	state	legislation	01/01/71	Provide footpaths and bike trails as part of road projects; minimum spending of 1 percent of city/county highway funds.	www.odot.state.or.us/techserv/bikewalk/plan_app/366514.htm
Pennsylvania Bicycle & Ped Checklist Training (Appendix J to PennDOT Design Manual)	state	manual, appendix	07/01/01	Developed as part of the statewide Bicycle and Pedestrian Master Plan, the "bicycle and pedestrian checklist" includes a comprehensive listing of the needs of pedestrians and cyclists that should be considered in appropriate transportation projects.	www.mail-archive.com/bike@list.purple.com/msg00613.html

Rhode Island state policy	state	legislation	06/19/97	Law says "department of transportation is authorized and directed to provide for the accommodation of bicycle and pedestrian traffic" design memo says "accommodations for bicyclists and pedestrians shall be considered."	www.rilin.state.ri.us/Statutes/TITLE31/31-18/31-18-21.HTM
South Carolina DOT Resolution	state	resolution, transportation commission	02/20/03	"…bicycling and walking accommodations should be a routine part of the Department's planning, design, construction and operating activities."	www.sccppa.org/advocacy/bike.html
Knoxville, Tennessee MPO Bicycle Accommodation Policy	MPO	plan	10/01/02	"Appropriate bicycle and pedestrian facilities shall be established in new construction and reconstruction projects in all urbanized areas unless one or more of three conditions are met."	www.knoxtrans.org/plans/bikeplan/index.htm
Tennessee DOT Bicycle and Pedestrian policy	state highways	internal policy; DOT directive	01/01/03	"The policy of TDOT is to routinely integrate bicycling and pedestrian facilities into the transportation system as a means to improve mobility and safety of non-motorized traffic."	www.tdot.state.tn.us/bikeroutes/policy.pdf
Virginia DOT Policy for Integrating Bicycle and Pedestrian Accommodations	state owned roads; jurisdiction over most county roads	internal policy	03/18/04	"The Virginia Department of Transportation (VDOT) will initiate all highway construction projects with the presumption that the projects shall accommodate bicycling and walking."	www.virginiadot.org/infoservice/news/newsrelease.asp?ID=CO-0414

APPENDIX E

Policy Examples

***Example 1: United States Department of Transportation Design Guidance
(Accommodating Bicycle and Pedestrian Travel)***

1. Bicycle and pedestrian ways shall be established in new construction and reconstruction projects in all urbanized areas unless one or more of three conditions are met:

   • Bicyclists and pedestrians are prohibited by law from using the roadway. In this instance, a greater effort may be necessary to accommodate bicyclists and pedestrians elsewhere within the right of way or within the same transportation corridor.

   • The cost of establishing bikeways or walkways would be excessively disproportionate to the need or probable use. Excessively disproportionate is defined as exceeding twenty percent of the cost of the larger transportation project.

   • Where scarcity of population or other factors indicate an absence of need. For example, the Portland Pedestrian Guide requires "all construction of new public streets" to include sidewalk improvements on both sides, unless the street is a cul-de-sac with four or fewer dwellings or the street has severe topographic or natural resource constraints.

2. In rural areas, paved shoulders should be included in all new construction and reconstruction projects on roadways used by more than 1,000 vehicles per day, as in States such as Wisconsin. Paved shoulders have safety and operational advantages for all road users in addition to providing a place for bicyclists and pedestrians to operate.

   Rumble strips are not recommended where shoulders are used by bicyclists unless there is a minimum clear path of four feet in which a bicycle may safely operate.

3. Sidewalks, shared use paths, street crossings (including over-and undercrossings), pedestrian signals, signs, street furniture, transit stops and facilities, and all connecting pathways shall be

designed, constructed, operated and maintained so that all pedestrians, including people with disabilities, can travel safely and independently.

4. The design and development of the transportation infrastructure shall improve conditions for bicycling and walking through the following additional steps:

    • Planning projects for the long-term. Transportation facilities are long-term investments that remain in place for many years. The design and construction of new facilities that meet the criteria in item 1) above should anticipate likely future demand for bicycling and walking facilities and not preclude the provision of future improvements. For example, a bridge that is likely to remain in place for 50 years, might be built with sufficient width for safe bicycle and pedestrian use in anticipation that facilities will be available at either end of the bridge even if that is not currently the case.

    • Addressing the need for bicyclists and pedestrians to cross corridors as well as travel along them. Even where bicyclists and pedestrians may not commonly use a particular travel corridor that is being improved or constructed, they will likely need to be able to cross that corridor safely and conveniently. Therefore, the design of intersections and interchanges shall accommodate bicyclists and pedestrians in a manner that is safe, accessible and convenient.

    • Getting exceptions approved at a senior level. Exceptions for the non-inclusion of bikeways and walkways shall be approved by a senior manager and be documented with supporting data that indicates the basis for the decision.

    • Designing facilities to the best currently available standards and guidelines. The design of facilities for bicyclists and pedestrians should follow design guidelines and standards that are commonly used, such as the AASHTO Guide for the Development of Bicycle Facilities, AASHTO's A Policy on Geometric Design of Highways and Streets, and the ITE recommended practice Design and Safety of Pedestrian Facilities.

Example 2: Mid-Ohio Regional Planning Commission Policy (Bicycle and Pedestrian Planning Policy, introductory section)

Many state, county and local jurisdictions are beginning to recognize the value and the need of routinely providing facilities for pedestrians or bicyclists. The inclusion of facilities in the early planning phases of new highway construction and residential and commercial development reduces the complexity and costs of attempting to retrofit years later. Mid-Ohio Regional Planning Commission (MORPC) encourages and supports those communities that have taken the step toward routinely accommodating pedestrians and bicyclists in the planning process. To others, MORPC encourages and supports the inclusion of routine accommodation by providing the following policy.

Project sponsors are required to accommodate bicycles and pedestrians in the planning and design of all proposed transportation projects using MORPC-attributable federal funds. Sponsors using local, state, or other federal funds are encouraged to accommodate bicycles and pedestrians in the planning and design of all proposed transportation projects. All transportation facilities on which bicyclists and pedestrians are permitted by law, including but not limited to streets, roads, highways, bridges, buses, trains, transit stops and facilities, and all connecting pathways shall be designed, constructed, operated and maintained so that all modes and pedestrians, including people with disabilities, can travel safely and independently.

Example 3: South Carolina Department of Transportation, Transportation Commission Resolution (on bicycling and walking)

RESOLUTION

WHEREAS, increasing walking and bicycling offers the potential for cleaner air, greater health of the population, reduced traffic congestion, more livable communities, less reliance on fossil fuels and their foreign supply sources and more efficient use of road space and resources; and

WHEREAS, in 2001 crashes involving bicyclists and pedestrians represented 13 percent of the traffic fatalities in S.C. and in the U.S.; and

WHEREAS, the Federal Highway Administration (FHWA) in its February 24, 1999 Policy statement "Guidance on the Bicycle and Pedestrian Provisions of the Federal-Aid Program" urges states to include bicycle and pedestrian accommodations routinely in their programmed highway projects; and

WHEREAS, bicycle and pedestrian projects and programs are eligible for funding from almost all of the major Federal-aid funding programs; and

WHEREAS, the South Carolina Department of Transportation Commission is strongly committed to improving conditions for walking and bicycling; and

WHEREAS, the Transportation Equity Act for the 21st Century (TEA-21) calls for the mainstreaming of bicycle and pedestrian projects into the planning, design and operation of our Nation's transportation system;

NOW, THEREFORE, BE IT RESOLVED that the South Carolina Department of Transportation Commission in meeting duly assembled this 14th day of January 2003, affirms that bicycling and walking accommodations should be a routine part of the department's planning, design, construction and operating activities, and will be included in the everyday operations of our transportation system; and

THEREFORE, BE IT FURTHER RESOLVED, that the South Carolina Department of Transportation Commission requires South Carolina counties and municipalities to make bicycling and pedestrian improvements an integral part of their transportation planning and programming where State or Federal Highway funding is utilized.

_____

L. Morgan Martin, Chairman

_____

Robert W. Harrell, First Congressional District

John N. Hardee, Second Congressional District

Eugene C. Stoddard, Third Congressional District

H. Howell Clyborne, Jr., Fourth Congressional District

B. Bayles Mack, Fifth Congressional District

John M. "Moot" Truluck, Sixth Congressional District

Appendix F

Complete Streets Policy Checklist

Pre-screen: Does the policy *require* that road projects be designed to accommodate all users? *If not, it does not qualify as a complete streets policy.*

1. Policy intent:
Is the policy part of a broader goal of providing a complete transportation network for all modes such as through the current strategic plan, transportation system upgrades, new administration's goals, etc.?

2. Policy Coverage:
2a. Does the policy cover motorists, bicyclists, pedestrians, transit users, and disabled users?
2b. Does the policy cover:
-all roads, regardless of responsible agency? (best)
OR:
-roads managed by single agency or roads seeking a specific funding source?
AND/OR:
-roads installed by private developers?

2c. Does the policy cover:
Construction? Reconstruction? Widenings? Other improvements? Repaving? Bridges? Stand-alone retrofit projects?

3. Policy requirements (beyond pre-screen requirement above):
When projects do not meet this standard, is there a formal process for approval of clearly stated exceptions placing the burden of proof on not accommodating all users?

4. Does the policy direct the use of the latest and best design standards?

5. Does the policy set performance standards?

6. Does the policy including a funding mechanism?

7. Implementation
Has the policy resulted in:
-restructured procedures?
-re-written design manuals or cross-sections?
-sessions for training planners and engineers?
-new data collection procedures?
-the creation of complete streets?

978-0-595-39318-3
0-595-39318-7

www.ingramcontent.com/pod-product-compliance
Lightning Source LLC
Chambersburg PA
CBHW080419290526
45791CB00008BA/2336

9 780595 393183